vol 3 of the series
Examples of Mathematical Structures

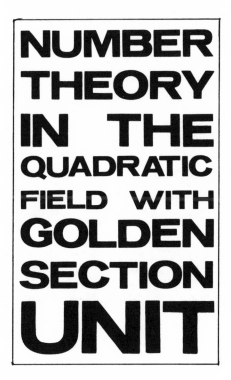

NUMBER THEORY IN THE QUADRATIC FIELD WITH GOLDEN SECTION UNIT

Fred Wayne Dodd

Polygonal Publishing House
80 Passaic Avenue
Passaic, NJ 07055

Library of Congress Cataloging in Publication Data

Dodd, Fred Wayne.
    Number theory in the quadratic field with golden
section unit.

        (Examples of mathematical structures ; v. 3)
        Bibliography: p.
        Includes index.
        1. Quadratic fields. 2. Numbers, Theory of. I. Title.
II. Series.
QA247.D66 1983   512'.74   83–17799
ISBN 0-936428-08-2

manufactured in the United States of America
by Braun-Brumfield

Contents

Preface

If D is the set of algebraic integers in a Euclidean quadratic number field, then it is possible to develop a number theory for D along the same lines as ordinary number theory; that is, number theory for $Z = \{0, \pm 1, \pm 2, \pm 3, \ldots \}$. This book is a case study of how such a development can be achieved for the algebraic integers in the Euclidean quadratic number field $Q(\sqrt{5})$. The reasons for choosing $Q(\sqrt{5})$ over other Euclidean quadratic number fields are discussed in Chapter 1. For now, suffice it to say, the number theory developed in this particular case has important applications to ordinary number theory, in particular, to the study of Fibonacci numbers.

I have endeavored to keep the presentation at an elementary level while at the same time, to use basic notions from abstract algebra to expedite portions of the theory and to cast the development in a more algebraic setting. Thus the prerequisites for reading this book are minimal. A rudimentary knowledge of groups, rings, and fields at the level of Herstein [10] is needed. In particular, it is assumed that the reader is well acquainted with the material on Euclidean rings found in Chapter 3 of that book. In addition, a first course in number theory at the level of Niven and Zuckerman [16] is assumed.

This book is an adaptation of my 1981 Doctor of Arts dissertation *Number Theory in the Integral Domain* $Z(\frac{1}{2} + \frac{1}{2}\sqrt{5})$. Chapters 1 and 4 have been revised and Appendix B has been greatly expanded.

Chapter 1 gives an introductory survey of quadratic number fields and the rationale for studying the algebraic integers in $Q(\sqrt{5})$. In Chapter 2 we commence the theory by showing that $Q(\sqrt{5})$ is a Euclidean quadratic number field and that the algebraic integers in $Q(\sqrt{5})$ are given by $Z(\omega) = \{c + d\omega : c, d \in Z\}$, where ω is the golden section $\frac{1}{2} + \frac{1}{2}\sqrt{5}$. In Chapter 3 the classification of the units and primes of $Z(\omega)$ is obtained, and the results are used to solve the Diophantine equation $x^2 + xy - y^2 = M$ and to obtain the prime factorization of a nonzero element of $Z(\omega)$. Chapter 4 is an introduction to discriminants and integral bases for ideals in $Z(\omega)$. Chapters 5–8 present the main theoretic components of the book. Congruence arithmetic in $Z(\omega)$ is developed. The results parallel those of ordinary number theory. Residue sets, Euler's totient function, the Chinese remainder theorem, the theorems of Euler and Fermat, linear and algebraic congruences, Wilson's theorem, primitive roots, indices, and Euler's criterion for nth power residues have analogues in $Z(\omega)$. The elements of $Z(\omega)$ with primitive roots are cataloged, and a theory of quadratic residues for $Z(\omega)$ is developed. The final chapter, Chapter 9, is devoted to number-theoretic applications of the theory. Fermat's last theorem for the case $n = 5$ is proved; and the Lucas test for primeness of the Mersenne number M_{4n+3} is verified. In the concluding section, the arithmetic theory of $Z(\omega)$ is used to deduce some of the well-known divisibility properties of the Fibonacci numbers.

I wish to thank my major adviser, Robert Tolar, and the other members of my dissertation committee, Donald Elliott, William Bosch, and Willard Fadner, for their positive contributions to this study. A very special thanks goes to Leon Mattics for his interest, contributions, and professional assistance. My wife, Fran, deserves special commendations for the many hours she spent improving the grammatical aspects of the final manuscript. I am indebted to George Hoffman for his assistance in preparing Appendix B. Finally, I am grateful to Michael Weinstein of Polygonal Publishing House for his encouragement and editorial assistance.

I Introduction

Purpose

If $\omega = \frac{1}{2}(1 + \sqrt{5})$ and $Z(\omega)$ denotes the set of all elements of the form $c + d\omega$, where c and d are elements of $Z = \{0, \pm 1, \pm 2, \ldots \}$, then $Z(\omega)$ under the usual operations of addition and multiplication of real numbers forms an integral domain. A number theory for $Z(\omega)$ can be developed which is similar in many respects to the number theory for Z. Furthermore, since $Z \subseteq Z(\omega)$, it is frequently possible to use the theory of $Z(\omega)$ to ascertain important results in the number theory for Z.

The purpose of this book is to present an elementary development of the number theory for $Z(\omega)$. The development will parallel the standard first course in number theory in which divisibility theory is emphasized. The concluding chapter of the study contains applications of the number theory for $Z(\omega)$ to three areas of interest—Mersenne numbers, Fermat's last theorem, and Fibonacci numbers.

Quadratic Number Fields

To further discuss the rationale for this study, it is necessary to consider the general concept of quadratic number fields. The following brief synopsis will suffice for the purposes in mind.

If Δ in $Z - \{0,1\}$ has no square factors greater than 1, and Q denotes the set of rational numbers, then $Q(\sqrt{\Delta})$ will denote the set of all numbers of the form $a + b\sqrt{\Delta}$, where a and b are in Q. Under the usual operations of addition and multiplication of complex numbers, $Q(\sqrt{\Delta})$ is a field and is referred to as a *quadratic number field*. If $\Delta < 0$, then $Q(\sqrt{\Delta})$ is called an *imaginary* quadratic number field, and if $\Delta > 0$, then $Q(\sqrt{\Delta})$ is called a *real* quadratic number field. We define

$$\sigma = \begin{cases} \sqrt{\Delta}, & \text{if } \Delta \equiv 2,3 \pmod 4 \\ \dfrac{1}{2} + \dfrac{1}{2}\sqrt{\Delta}, & \text{if } \Delta \equiv 1 \pmod 4 \end{cases}$$

and denote the set of all numbers of the form $c + d\sigma$, where c and d are in Z, by $Z(\sigma)$. Each element of $Z(\sigma)$ is called a *quadratic integer* of the field $Q(\sqrt{\Delta})$. Following the convention of algebraic number theorists, we shall simply call elements of $Z(\sigma)$ *integers* and elements of Z *rational integers*. The terminology is apt, because the only elements of $Q(\sqrt{\Delta}) \cap Q$ which are quadratic integers are elements of Z. In the same vein, we call the ordinary primes of Z *rational primes,* and the number theory of Z will be called *rational integer number theory*. It is quite elementary to verify that $Z(\sigma)$ is an integral domain with identity 1. It should be noted that the integral domain $Z(\omega)$, introduced earlier, is the special case $\Delta = 5$ and $\sigma = \omega$.

Since $Z(\sigma)$ is an integral domain with identity 1, then the standard terminology of units, associates, primes, and unique factorization domains is applicable. But it must be noted that our primes are what many algebraists call irreducible elements; that is, a nonunit π in $Z(\sigma) - \{0\}$ is a *prime* if whenever $\pi = \alpha\beta$, where α and β are in $Z(\sigma)$, then either α or β is a unit. If $Q(\sqrt{\Delta})$ is an imaginary quadratic number field, then $Z(\sigma)$ has only a finite number of units; and if $Q(\sqrt{\Delta})$ is a real quadratic number field, then $Z(\sigma)$ has an infinite number of units. The existence of an infinite number of units adds complications, so that, in a sense, the study of number theory in real quadratic number fields is more difficult than in imaginary quadratic number fields. It can be shown that any nonunit α in $Z(\sigma) - \{0\}$ is expressible as a finite product of primes of $Z(\sigma)$. Whenever $Z(\sigma)$ has the unique factorization property, then much of rational integer number theory can be extended to $Z(\sigma)$. Unfortunately, unique factorization domains are the exception rather than the rule in quadratic number fields.

The imaginary quadratic number fields $Q(\sqrt{\Delta})$ for which $Z(\sigma)$ is a unique factorization domain have been completely characterized. They are $Q(\sqrt{\Delta})$, where $\Delta = -1, -2, -3, -7, -11, -19, -43, -67$, and -163 (Stark [19], p. 295). It has been conjectured that there are an infinite number of real quadratic number fields for which $Z(\sigma)$ is a unique factorization domain. This conjecture has not been proved (Stark [19], p. 296).

If $\alpha = a + b\sqrt{\Delta}$, where a and b are in Q, then $\bar{\alpha} = a - b\sqrt{\Delta}$ is called the *conjugate* of α and $N(\alpha) = \alpha\bar{\alpha}$ is called the *norm* of α. When α is in $Z(\sigma) - \{0\}$, then $N(\alpha)$ is a nonzero rational integer. $Q(\sqrt{\Delta})$ is called a *Euclidean* quadratic number field if for every α and β in $Z(\sigma)$ with $\beta \neq 0$, there exist θ and ρ in $Z(\sigma)$ such that $\alpha = \beta\theta + \rho$ and $|N(\rho)| < |N(\beta)|$. If $Q(\sqrt{\Delta})$ is a Euclidean quadratic number field, then $Z(\sigma)$ is a unique factorization domain. Euclidean quadratic number fields have been completely determined. They are $Q(\sqrt{\Delta})$, where $\Delta = -1, -2, -3, -7, -11, 2, 3, 5, 6, 7, 11, 13, 17, 19, 21, 29, 33, 37, 41, 57,$ and 73 (Hardy and Wright [9], p. 213).

The rational integers are contained in any $Z(\sigma)$. Thus for a given σ we have, in a sense, enriched the set of rational integers. The significance of this enrichment is that many number-theoretic problems concerning the rational integers, that would otherwise be difficult or impossible to solve within the more restricted framework of Z, can be solved within the richer framework of $Z(\sigma)$.

Historical Background

Fermat's last theorem states that the equation $x^n + y^n = z^n$ has no solution in positive rational integers for $n > 2$. Proofs of this theorem for the particular cases $n = 3$ and $n = 5$ represent two of the earliest applications of quadratic number fields to rational integer number theory. In 1770 Euler used the integers of $Z(\frac{1}{2} + \frac{1}{2}\sqrt{-3})$ in giving a proof of Fermat's last theorem for $n = 3$ (see Edwards [8], pp. 39–54). Then in 1825 Dirichlet and Legendre used $Z(\omega)$ in proving the theorem for the case $n = 5$ (Edwards [8], pp. 65–73). However, it was Gauss who made the first really significant use of quadratic number fields, when around 1830 he developed the number theory of $Z(\sqrt{-1})$ in his researches on biquadratic residues (Reid [18], pp. 205–217). (If a and m are relatively prime rational integers, then a is said to be a *biquadratic residue* or *nonresidue* of m according as the congruence $x^4 \equiv a$ (mod m) has or has not solutions.) Today, the integers in $Z(\sqrt{-1})$ are called *Gaussian integers* in honor of Gauss and his pioneer work in quadratic number fields. Gauss's favorite pupil, Eisenstein, used $Z(\frac{1}{2} + \frac{1}{2}\sqrt{-3})$ to obtain number-theoretic results concerning cubic residues (Hardy and Wright [9], p. 189).

The theory of quadratic number fields was greatly extended by Kummer (1810–1892), Dedekind (1831–1916), and others to number fields of higher dimensions. Their work forms the basis of an important branch of mathematics known as algebraic number theory. Kummer's work in algebraic number theory allowed him to make the most successful attack on Fermat's last theorem to date, and many of the methods he used have been incorporated into the mainstream of modern mathematics, as, for example, ideal theory. An excellent elementary account of this early history of algebraic number theory can be found in Edwards [7]. A much deeper investigation can be found in Edwards [8].

Despite this rich history of quadratic number fields, significant advances are still being made. The most notable was in 1966, when H. M. Stark completed the task of cataloging the nine imaginary quadratic fields $Q(\sqrt{\Delta})$ for which $Z(\sigma)$ is a unique factorization domain (Stark [19], pp. 295–296).

$Z(\omega)$ and Rationale for Its Study

As previously noted, the elements of $Z(\omega)$ are the integers of the quadratic number field $Q(\sqrt{5})$. The integer ω is of fundamental importance, since it can be shown that ϵ in $Z(\omega)$ is a unit if and only if $\epsilon = \pm \omega^n$ for some n in Z (see Theorem 3.1). For this reason ω is called the *fundamental unit* of $Z(\omega)$. The integer ω is an intriguing number. It occurs frequently in geometry and other areas of mathematics and is usually called the *divine proportion* or *golden section*. The ancients ascribed aesthetic and mystical qualities to geometric figures in which the golden section was found. Thus, for example, a rectangle whose sides are in the ration $\omega{:}1$ is called a *golden rectangle* and supposedly has a pleasant appearance to the eye.

My interest in $Z(\omega)$ began when it was brought to my attention by Leon Mattics of the University of South Alabama that $Z(\omega)$ was the proper domain in which to study the arithmetic properties of the Fibonacci sequence 1, 1, 2, 3, 5, 8, 13, 21, . . . , and the Lucas sequence 1, 3, 4, 7, 11, 18, 29, 47, ... This is clearly the case, since the nth terms F_n and L_n, of these two sequences can be expressed in terms of the fundamental unit ω. Specifically, $F_n = (\omega^n - \overline{\omega}^n)/(\omega - \overline{\omega})$ and $L_n = \omega^n + \overline{\omega}^n$.

Amongst the quadratic number fields, $Q(\sqrt{-1})$ has received the most individual attention. This emphasis is warranted because of the applications of $Z(\sqrt{-1})$ to Pythagorean triples, sums of squares, and, as previously mentioned, the study of biquadratic residues. In addition, $Q(\sqrt{-1})$ is a Euclidean number field, and $Z(\sqrt{-1})$ thus serves as an excellent illustration of how number theory can be developed along lines similar to that of rational integer theory.

Among the real Euclidean quadratic number fields, I feel that $Q(\sqrt{5})$ is the most important and should be singled out for special emphasis much as $Q(\sqrt{-1})$ has been singled out. The reasons for this belief have been alluded to previously, but a succinct synopsis follows: $Q(\sqrt{5})$ is a Euclidean number field, and so a number theory for $Z(\omega)$ can be constructed that will parallel that of rational integer theory. The presence in $Z(\omega)$ of an infinite number of units can be expected to lead to difficulties not encountered in the Gaussian integers. Historically, $Z(\omega)$ was used to prove Fermat's last theorem for $n = 5$ and appears to have been the first real quadratic number field to have been applied to rational integer number theory. Finally, the close connection of ω with the sequences of Fibonacci and Lucas may be exploited as an aid in developing the arithmetic theory of $Z(\omega)$; conversely, the theory of $Z(\omega)$ can be used to ascertain arithmetic properties of these sequences.

References to Rational Integer Number Theory

Since the purpose of this study is to develop a number theory for $Z(\omega)$ along the lines of rational integer number theory, it follows that much of the theory has been obtained by suitably modifying the corresponding rational integer theory. The principal sources for the requisite rational integer theory are Hardy and Wright [9], Hunter [11], and Niven and Zuckerman [16]. In particular, the work by Hunter deserves special citation, for much of the material in Chapters 5–8 has been patterned along lines found in his book.

II   Elementary Divisibility Properties of Z(ω)

Introduction

In this chapter we commence the development of number theory in $Z(\omega)$ by investigating some of its divisibility properties. As a reminder we emphasize that the symbol ω is reserved exclusively to denote the number $\frac{1}{2} + \frac{1}{2}\sqrt{5}$ and $Z(\omega)$ denotes the set of numbers $c + d\omega$, where c and d are in Z. The main result will be the establishment of $Q(\sqrt{5})$ as a Euclidean quadratic number field; thus $Z(\omega)$ is a Euclidean domain, and so has the divisibility properties common to all Euclidean domains.

Arithmetic in $Q(\sqrt{5})$

We recall that $Q(\sqrt{5})$ is the smallest field that contains Q and $\sqrt{5}$ and that $Q(\sqrt{5})$ is a vector space of dimension 2 over Q. Thus, if α and β in $Q(\sqrt{5})$ are linearly independent over Q, then $\{\alpha, \beta\}$ is a basis for $Q(\sqrt{5})$ over Q. Consequently, every element of $Q(\sqrt{5})$ can be uniquely expressed in the form $a\alpha + b\beta$, where a and b are in Q. Specific bases of interest are $\{1, \sqrt{5}\}$ and $\{1, \omega\}$. For reasons that will become apparent in Theorem 2.6, it is convenient to represent an arbitrary α in $Q(\sqrt{5})$ in the form $\alpha = \frac{1}{2}a + \frac{1}{2}b\sqrt{5}$ when using the basis $\{1, \sqrt{5}\}$. Our first theorem gives the relationship between the bases $\{1, \sqrt{5}\}$ and $\{1, \omega\}$.

THEOREM 2.1 *If a, b, c, and d are rational numbers and $\frac{1}{2}a + \frac{1}{2}b\sqrt{5} = c + d\omega$, then $a = 2c + d$, $b = d$, and $c = \frac{1}{2}(a - b)$.*
PROOF: Since $c + d\omega = c + \frac{1}{2}d(\sqrt{5} + 1) = \frac{1}{2}(2c + d) + \frac{1}{2}d\sqrt{5}$, then $a = 2c + d$ and $b = d$. But then $a = 2c + b$ and whence the result $c = \frac{1}{2}(a - b)$. Q.E.D.

Since for the most part we will be using the basis $\{1, \omega\}$, the next theorem is indispensable when doing multiplicative arithmetic in $Q(\sqrt{5})$.

THEOREM 2.2 $\omega^2 = \omega + 1$.
Proof: We have $\omega^2 = \frac{1}{4}(1 + \sqrt{5})^2 = \frac{1}{4}(6 + 2\sqrt{5}) = \frac{1}{2}(1 + \sqrt{5}) + 1 = \omega + 1$. Q.E.D.

EXAMPLE 2.1 Using Theorem 2,2, we obtain $(1 + 2\omega)(3 + 5\omega) = 3 + 6\omega + 5\omega + 10\omega^2 = 3 + 11\omega + 10\omega + 10 = 13 + 21\omega$.

The conjugate and norm functions were defined for quadratic number fields in Chapter 1. We repeat the definitions here.

DEFINITIONS Let a and b be rational numbers and $\alpha = \frac{1}{2}a + \frac{1}{2}b\sqrt{5}$. The **conjugate** of α, denoted by $\bar{\alpha}$, is given by $\bar{\alpha} = \frac{1}{2}a - \frac{1}{2}b\sqrt{5}$. The **norm** of α, denoted by $N(\alpha)$, is given by $N(\alpha) = \alpha\bar{\alpha} = \frac{1}{4}(a^2 - 5b^2)$.

The remaining theorems of this section list the basic properties of the conjugate and norm functions.

THEOREM 2.3

(i) $\bar{\omega} = 1 - \omega$.

(ii) $N(\omega) = \omega\bar{\omega} = -1$.

(iii) $\bar{\omega}^2 = \bar{\omega} + 1$.

Proof: Assertion (i) follows from $\omega + \bar{\omega} = (\frac{1}{2} + \frac{1}{2}\sqrt{5}) + (\frac{1}{2} - \frac{1}{2}\sqrt{5}) = 1$, and (ii) follows from $\omega\bar{\omega} = \frac{1}{4}(1 + \sqrt{5})(1 - \sqrt{5}) = \frac{1}{4}(1 - 5) = -1$. As for (iii), we note that $\bar{\omega}^2 = (1 - \omega)^2 = 1 - 2\omega + \omega^2 = 1 - 2\omega + \omega + 1 = 1 - \omega + 1 = \bar{\omega} + 1$. Q.E.D.

THEOREM 2.4 *Let $\alpha = c + d\omega$, where c and d are rational numbers. Then $\bar{\alpha} = c + d\bar{\omega}$ and $N(\alpha) = c^2 + cd - d^2$.*

PROOF: Using the notation and content of Theorem 2.1, we obtain $c + d\bar{\omega} = \frac{1}{2}(a - b) + \frac{1}{2}b(1 - \sqrt{5}) = \frac{1}{2}a - \frac{1}{2}b\sqrt{5} = \bar{\alpha}$. Then, by Theorem 2.3, we deduce that $N(\alpha) = \alpha\bar{\alpha} = (c + d\omega)(c + d\bar{\omega}) = c^2 + cd(\omega + \bar{\omega}) + d^2\omega\bar{\omega} = c^2 + cd - d^2$. Q.E.D.

THEOREM 2.5 *If α and β are in $Q(\sqrt{5})$, then*

(i) $\overline{(\bar{\alpha})} = \alpha$,

(ii) $\bar{\alpha} = \alpha$ *if and only if α is a rational number,*

(iii) $N(\alpha) = 0$ *if and only if $\alpha = 0$,*

(iv) $N(\alpha)$ *and $\alpha + \bar{\alpha}$ are rational numbers,*

(v) α *and $\bar{\alpha}$ are the solutions of $X^2 - (\alpha + \bar{\alpha})X + N(\alpha) = 0$,*

(vi) $\overline{\alpha \pm \beta} = \bar{\alpha} \pm \bar{\beta}$,

(vii) $\overline{\alpha\beta} = \bar{\alpha}\bar{\beta}$,

(viii) $\overline{(\alpha/\beta)} = \bar{\alpha}/\bar{\beta}$ *when $\beta \neq 0$,*

(ix) $N(\alpha\beta) = N(\alpha)N(\beta)$,

(x) $N(\alpha/\beta) = N(\alpha)/N(\beta)$ *when $\beta \neq 0$.*

PROOF: The truth of the first two assertions is clear from the defintion of conjugate. Since $N(\alpha) = \alpha\bar{\alpha}$, then $N(\alpha) = 0$ if and only if $\alpha = 0$ or $\bar{\alpha} = 0$; but clearly $\bar{\alpha} = 0$ if and only if $\alpha = 0$. Accordingly, $N(\alpha) = 0$ if and only if $\alpha = 0$, and (iii) is true. Assertion (iv) is obvious from the definitions of norm and conjugate; and (v) follows from $(X - \alpha)(X - \bar{\alpha}) = X^2 - (\alpha + \bar{\alpha})X + N(\alpha)$.

In order to establish (vi)–(viii), set $\alpha = c + d\omega$ and $\beta = r + s\omega$, where c, d, r, and s are rational numbers. Then, from Theorem 2.4, we have

$$\overline{\alpha \pm \beta} = \overline{(c \pm r) + (d \pm s)\omega} = (c \pm r) + (d \pm s)\overline{\omega}$$

$$= (c + d\overline{\omega}) \pm (r + s\overline{\omega}) = \overline{\alpha} \pm \overline{\beta},$$

so that (vi) is true.

A simple calculation yields that $\alpha\beta = (c + d\omega)(r + s\omega) = cr + ds + (cs + dr + ds)\omega$, so that $\overline{\alpha\beta} = cr + ds + (cs + dr + ds)\overline{\omega}$. Also, with the aid of Theorem 2.3(iii), we obtain $\overline{\alpha}\overline{\beta} = (c + d\overline{\omega})(r + s\overline{\omega}) = cr + ds + (cs + dr + ds)\overline{\omega}$, and so we deduce that $\overline{\alpha\beta} = \overline{\alpha}\overline{\beta}$. Thus (vii) is true. Since $t = N(\beta) = \beta\overline{\beta}$ is rational and $1/\beta = \overline{\beta}/\beta\overline{\beta} = r/t + (s/t)\overline{\omega}$, then $\overline{(1/\beta)} = r/t + (s/t)\omega$. But we also have $1/\overline{\beta} = \beta/\beta\overline{\beta} = r/t + (s/t)\omega$, and thus $\overline{(1/\beta)} = 1/\overline{\beta}$. Consequently, we deduce from (vii) that $\overline{(\alpha/\beta)} = \overline{\alpha(1/\beta)} = \overline{\alpha}\overline{(1/\beta)} = \overline{\alpha}(1/\overline{\beta}) = \overline{\alpha}/\overline{\beta}$. Hence (viii) is true. Next we note from (vii) that $N(\alpha\beta) = (\alpha\beta)\overline{\alpha\beta} = (\alpha\beta)(\overline{\alpha}\overline{\beta}) = (\alpha\overline{\alpha}) \cdot (\beta\overline{\beta}) = N(\alpha)N(\beta)$. This establishes (ix). Finally, we conclude from (viii) that $N(\alpha/\beta) = (\alpha/\beta)\overline{(\alpha/\beta)} = (\alpha/\beta)(\overline{\alpha}/\overline{\beta}) = (\alpha\overline{\alpha})/(\beta\overline{\beta}) = N(\alpha)/N(\beta)$. Thus (x) is true, and the proof is complete. Q.E.D.

Quotient arithmetic in $Q(\sqrt{5})$ is reminiscent of complex arithmetic. We conclude the section with a simple example.

EXAMPLE 2.2

$$\frac{3 + 5\omega}{2 + \omega} = \frac{(3 + 5\omega)(2 + \overline{\omega})}{(2 + \omega)(2 + \overline{\omega})} = \frac{6 + 10\omega + 3\overline{\omega} + 5\omega\overline{\omega}}{N(2 + \omega)}$$

$$= \frac{6 + 10\omega + 3(1 - \omega) - 5}{4 + 2 - 1} = \frac{4}{5} + \frac{7}{5}\omega.$$

Integers of $Q(\sqrt{5})$

We use the terminology given in Chapter 1. Thus the integers of $Q(\sqrt{5})$ are the elements of $Z(\omega)$, the elements of Z are called rational integers, the ordinary primes of Z are called rational primes, and the number theory of Z is called rational integer number theory. It should be noted that this terminology contains a subtle ambiguity. For while it is true that a rational integer is an integer (an element of $Z(\omega)$), it is not true, as we shall discover in the next chapter, that a rational prime is necessarily a prime of $Z(\omega)$.

For the remainder of the book we will have no occasion to consider quadratic fields other than $Q(\sqrt{5})$. Accordingly, the term integer will henceforth mean an element of $Z(\omega)$. Also, in the chapters to follow, adherence to the notation $Z(\omega)$ is inconvenient, and so we adopt the following less suggestive but more economical notation:

NOTATION $\hat{Z} = Z(\omega)$.

The next theorem characterizes the elements of $\hat{Z}$ in terms of the basis $\{1,\sqrt{5}\}$ of $Q(\sqrt{5})$.

THEOREM 2.6 *The element α in $Q(\sqrt{5})$ is an integer if and only if $\alpha = \frac{1}{2}a + \frac{1}{2}b\sqrt{5}$ for some rational integers a and b with the same parity.*

PROOF: If α is an integer, then $\alpha = c + d\omega = c + \frac{1}{2}(1 + \sqrt{5})d$ $= \frac{1}{2}(2c + d) + \frac{1}{2}d\sqrt{5}$, and it is obvious that $a = 2c + d$ and $b = d$ are rational integers with the same parity.

Conversely, suppose that a and b are rational integers with the same parity and $\alpha = \frac{1}{2}a + \frac{1}{2}b\sqrt{5}$. Then, by Theorem 2.1, $\alpha = c + d\omega$, where $c = \frac{1}{2}(a - b)$ and $d = b$. Furthermore, since a and b have the same parity, then both c and d are integral. Therefore α is an integer. Q.E.D.

If $\alpha = c + d\omega$ and $\beta = r + s\omega$, then $\alpha\beta = cr + ds + (cs + dr + ds)\omega$. Thus, if α and β are integers, then $\alpha\beta$ is an integer. It is trivial to verify that $\hat{Z}$ satisfies the other properties of an integral domain. We record this as

THEOREM 2.7 *$\hat{Z}$ is an integral domain with identity element 1.*

We conclude this section by showing that the definition of integer is appropriate in the sense that the set of algebraic integers in $Q(\sqrt{5})$ is precisely $\hat{Z}$. We recall the following basic information on algebraic numbers and algebraic integers (Niven and Zuckerman [16], pp. 235–237). A complex number Δ is said to be an *algebraic number* if it is a root of a polynomial $f(X)$ having rational coefficients. If Δ is an algebraic number, then it is a root of a unique monic polynomial $m_\Delta(X)$ which has rational coefficients and is irreducible over Q. Moreover, $m_\Delta(X)$ divides every polynomial $f(X)$ having rational coefficients and root Δ. An algebraic number Δ is said to be an *algebraic integer* if $m_\Delta(X)$ has rational integer coefficients.

Armed with these recollections, we proceed to show that every element of $Q(\sqrt{5})$ is an algebraic number and that the set of algebraic integers of $Q(\sqrt{5})$ is precisely $\hat{Z}$. According to Theorem 2.5, any element α in $Q(\sqrt{5})$ is a root of the polynomial $f(X) = X^2 - (\alpha + \bar{\alpha})X + N(\alpha)$ and, by the same theorem $f(X)$ has rational coefficients. Thus any α in $Q(\sqrt{5})$ is an algebraic number.

If α in $\hat{Z}$ is a rational integer, then α is an algebraic integer because, in this case $m_\alpha(X) = X - \alpha$. If α in $\hat{Z}$ is not a rational integer, then $\alpha = c + d\omega$, where c and d are in Z and $d \neq 0$. Consequently, α is irrational, and so $m_\alpha(X) = X^2 - (\alpha + \bar{\alpha})X + N(\alpha) = X^2 - (2c + d)X + c^2 + dc - d^2$. Thus $m_\alpha(X)$ has rational integer coefficients, and so α is an algebraic integer.

At this point we have shown that every element of $Q(\sqrt{5})$ is an algebraic number and that every element of $\hat{Z}$ is an algebraic integer. To complete the presentation, we must show that every algebraic integer in $Q(\sqrt{5})$ is an element of $\hat{Z}$. Therefore suppose that α in $Q(\sqrt{5})$ is an algebraic integer. Since α is in $Q(\sqrt{5})$, then we may write α in the form $\alpha = \frac{1}{2}a + \frac{1}{2}b\sqrt{5}$, where a and b are rational numbers. If $b = 0$, then α is rational, and so $m_\alpha(X) = X - \alpha$. But since α is an algebraic integer, then $m_\alpha(X)$ has rational integer coefficients; and so α is a rational integer (and thus an element of $\hat{Z}$). Hence we may suppose that $b \neq 0$. Then α is irrational, and so $m_\alpha(X) = X^2 - (\alpha + \bar{\alpha})X + N(\alpha) = X^2 - aX + \frac{1}{4}(a^2 - 5b^2)$. Since α is an algebraic integer, then a and

$\frac{1}{4}(a^2 - 5b^2) = n$ are rational integers. The latter equation can be written $5b^2 = a^2 - 4n$, so b must also be a rational integer, for otherwise, 5 being square-free, would't be able to clear the denominator of b^2. Thus a and b are rational integers, and $b^2 \equiv 5b^2 \equiv a^2$ (mod 4). But $a^2 \equiv b^2$ (mod 4) if and only if a and b have the same parity. Consequently, $\alpha = \frac{1}{2}a + \frac{1}{2}b\sqrt{5}$, where a and b are rational integers of the same parity. Thus it follows from Theorem 2.6 that α is in $\hat{Z}$, and our presentation is complete.

Conventions

The principal sets under consideration for the remainder of the book will be Z and $\hat{Z}$. For brevity of exposition we therefore adopt the following conventions:

CONVENTION 2.1   A number represented by a lowercase Latin letter is a rational integer unless it is specifically defined as an element of the more general set Q.

CONVENTION 2.2   A number represented by a lowercase Greek letter is an integer unless it is specifically defined as an element of the more general set $Q(\sqrt{5})$.

Division Algorithm for $\hat{Z}$

The integers have a division algorithm. Specifically, given α and β with $\beta \neq 0$, there exist θ and ρ such that $\alpha = \beta\theta + \rho$ and $|N(\rho)| < |N(\beta)|$. The following lemma paves the way for this result.

LEMMA 2.8 *Given any* Δ *in* $Q(\sqrt{5})$, *there exists a* θ *such that* $|N(\Delta - \theta)| \le 3/4$.

PROOF: Let $\Delta = r + s\omega$, where r and s are rational numbers. Then there exist rational integers c and d such that $|r - c| \le \frac{1}{2}$ and $|s - d| \le \frac{1}{2}$. If we set $\theta = c + d\omega$, then we have $|N(\Delta - \theta)| = |N((r - c) + (s - d)\omega| = |(r - c)^2 + (r - c)(s - d) - (s - d)^2| \le \frac{1}{4} + \frac{1}{4} + \frac{1}{4} = \frac{1}{4}(3)$. Q.E.D.

As promised, we now have

THEOREM 2.9 (Division algorithm for $\hat{Z}$) *Given* α *and* β *with* $\beta \ne 0$, *there exist integers* θ *and* ρ *such that* $\alpha = \beta\theta + \rho$ *and* $|N(\rho)| < |N(\beta)|$. *The integer* θ *is called the* quotient, *and the integer* ρ *is called the* remainder.

PROOF: Since α/β is in $Q(\sqrt{5})$, it follows from Lemma 2.8 that $|N(\alpha/\beta - \theta)| < 1$ for some θ. Hence $|N(\alpha - \beta\theta)| < |N(\beta)|$. If we set $\rho = \alpha - \beta\theta$, then $\alpha = \beta\theta + \rho$ and $|N(\rho)| < |N(\beta)|$. Q.E.D.

Actually, the stronger statement $|N(\rho)| \le (3/4)|N(\beta)|$ could have been used in the preceding theorem, but the form given will be adequate for our purposes. The proofs of Lemma 2.8 and Theorem 2.9 give a thoroughly practical method for determining θ and ρ. The following example illustrates the technique.

EXAMPLE 2.3 We shall find integers θ and ρ such that $2 = (-1 + 2\omega)\theta + \rho$ and $|N(\rho)| < |N(-1 + 2\omega)| = 5$. Since $2/(-1 + 2\omega) = -2/5 + (4/5)\omega$, $|-2/5 - 0| \le 1/2$, and $|4/5 - 1| \le 1/2$, then $\theta = 0 + 1\omega = \omega$ and $\rho = 2 - (-1 + 2\omega)\omega = -\omega$ will suffice; that is, $2 = (-1 + 2\omega)\omega - \omega$ and $|N(-\omega)| < |N(-1 + 2\omega)|$.

It should be noted that the value for ρ given in Example 2.3 is not the only possible integer which satisfies the conclusion of Theorem 2.9. For example, $2 = (-1 + 2\omega)(0) + 2$, $|N(2)| < |N(-1 + 2\omega)|$ and $2 = (-1 + 2\omega)(-1 + \omega) - 1 + \omega$, $|N(-1 + \omega)| < |N(-1 + 2\omega)|$. There are, in fact, an infinite number of different integers ρ such that $|N(\rho)| < 5$ and $2 = (-1 + 2\omega)\theta + \rho$ for some θ depending on ρ. This fact will be substantiated in the last section of Chapter 5.

$\hat{Z}$ is a Euclidean Domain

The division algorithm for $\hat{Z}$ has laid the groundwork for establishing $Q(\sqrt{5})$ as a Euclidean quadratic number field, or, what is the same thing, showing that $\hat{Z}$ is a Euclidean domain. We shall need the following simple result:

THEOREM 2.10 *If α is an integer, then $N(\alpha)$ is a rational integer and $N(\alpha) = 0$ if and only if $\alpha = 0$. Moreover, if β is a nonzero integer, then $|N(\alpha)| \le |N(\alpha\beta)|$.*
PROOF: It follows from Theorem 2.4 that $N(\alpha)$ is a rational integer and from Theorem 2.5(iii) that $N(\alpha) = 0$ if and only if $\alpha = 0$. Thus, since $|N(\beta)|$ is a positive rational integer, we get that $|N(\alpha\beta)| = |N(\alpha)N(\beta)| = |N(\alpha)| \cdot |N(\beta)| \ge |N(\alpha)|$. Q.E.D.

THEOREM 2.11 $\hat{Z}$ *is a Euclidean domain.*
PROOF: We need only exhibit a mapping T from $\hat{Z} - \{0\}$ into the nonnegative rational integers with the following two properties:
   (i) $T(\alpha) \le T(\alpha\beta)$ for all nonzero α and β.
   (ii) For any α and β with $\beta \ne 0$, there exist θ and ρ such that $\alpha = \beta\theta + \rho$, where either $\rho = 0$ or $T(\rho) < T(\beta)$.
We set $T(\alpha) = |N(\alpha)|$. It then follows from Theorems 2.9 and 2.10 that T satisfies Properties (i) and (ii). Q.E.D.

Euclidean Domain Properties

Now that we have established that $\hat{Z}$ is a Euclidean domain, we may apply the theory of Euclidean domains to $\hat{Z}$. Since this theory will be of fundamental importance for the remaining chapters, we conclude this chapter by cataloging the results we shall need. Verifications of most of the results of this section can be found in Herstein [10], pp. 143–149.

THEOREM 2.12   *Every ideal of $\hat{Z}$ is a principal ideal; that is, if I is an ideal of $\hat{Z}$, then there exists a μ such that $I = \mu\hat{Z} = \{ \mu\alpha : \alpha \in \hat{Z} \}$.*

DEFINITION   If $\alpha \neq 0$, then α is said to **divide** β if there exists a γ such that $\beta = \alpha\gamma$. We shall use the symbol $\alpha \mid \beta$ to represent the fact that α divides β and $\alpha \nmid \beta$ to mean that α does not divide β.

THEOREM 2.13
(i) *If $\alpha \mid \beta$ and $\beta \mid \gamma$, then $\alpha \mid \gamma$.*
(ii) *If $\alpha \mid \beta$ and $\alpha \mid \gamma$, then $\alpha \mid (\beta \pm \gamma)$.*
(iii) *If $\alpha \mid \beta$, then $\alpha \mid \beta\xi$ for any ξ.*

DEFINITION   The integer ϵ is a **unit** in $\hat{Z}$ if $\epsilon \mid 1$.

THEOREM 2.14   *The units in $\hat{Z}$ form a multiplicative abelian group. Thus, if ϵ_1 and ϵ_2 are units in $\hat{Z}$, then so are $\epsilon_1\epsilon_2$ and $\epsilon_1^{-1} = 1/\epsilon_1$.*

THEOREM 2.15   *If $\alpha \mid \beta$ and $\beta \mid \alpha$, then $\alpha = \epsilon\beta$ for some unit ϵ.*

THEOREM 2.16   *The integer α is a unit if and only if $|N(\alpha)| = 1$.*

DEFINITION   The integers α and β are said to be **associates** if $\alpha = \epsilon\beta$ for some unit ϵ.

THEOREM 2.17   *The relation of being associates is an equivalence relation on $\hat{Z}$. Thus*
(i) *α and α are associates;*
(ii) *if α and β are associates, then β and α are associates;*
(iii) *if α and β are associates and β and γ are associates, then α and γ are associates.*

DEFINITION   If not both α and β are zero, then δ is said to be a **greatest common divisor** of α and β if
(i) $\delta \mid \alpha$ and $\delta \mid \beta$;
(ii) whenever $\gamma \mid \alpha$ and $\gamma \mid \beta$, then $\gamma \mid \delta$.
We shall use the notation $\delta = \text{GCD}(\alpha, \beta)$ to denote that δ is a greatest common divisor of α and β.

THEOREM 2.18 *If α and β are not both zero, then*
(i) *α and β have a greatest common divisor δ;*
(ii) *δ' is a greatest common divisor of α and β if and only if δ and δ' are associates;*
(iii) *$\delta = \lambda\alpha + \kappa\beta$ for some λ and κ;*
(iv) *$\mathrm{GCD}(\alpha, \beta) = \mathrm{GCD}(\alpha + \gamma\beta, \beta)$.*

It should be noted that the equality in Theorem 2.18(iv) only holds up to associates; that is, if δ is a greatest common divisor of α and β, and δ' is a greatest common divisor of $\alpha + \gamma\beta$ and β, then δ and δ' are associates. We shall frequently use this license of notation.

The division algorithm may be used in the usual fashion to compute $\mathrm{GCD}(\alpha_1, \alpha_2)$:

$$\alpha_1 = \alpha_2\theta_1 + \alpha_3, \text{ where } |\mathrm{N}(\alpha_3)| < |\mathrm{N}(\alpha_2)|.$$

$$\alpha_2 = \alpha_3\theta_2 + \alpha_4, \text{ where } |\mathrm{N}(\alpha_4)| < |\mathrm{N}(\alpha_3)|.$$

$$\vdots$$

$$\alpha_{r-2} = \alpha_{r-1}\theta_{r-2} + \alpha_r, \text{ where } |\mathrm{N}(\alpha_r)| < |\mathrm{N}(\alpha_{r-1})|.$$

$$\alpha_{r-1} = \alpha_r\theta_{r-1}, \text{ and } \alpha_r = \mathrm{GCD}(\alpha_1, \alpha_2).$$

We shall use the notation $\gcd(a, b)$ in the sense of rational integer number theory; that is, $\gcd(a, b)$ is the unique largest positive rational integer which divides both a and b. It is evident from the preceding algorithm that $\gcd(a, b) = \mathrm{GCD}(a, b)$.

EXAMPLE 2.4 We find a greatest common divisor δ of $95 + 4\omega$ and $71 - 16\omega$, as well as λ and κ such that $\delta = \lambda(95 + 4\omega) + \kappa(71 - 16\omega)$. Using the technique found in Example 2.3, we deduce:

$$95 + 4\omega = (71 - 16\omega)(1) + (24 + 20\omega),$$

where $|\mathrm{N}(24 + 20\omega)| = 656 < 3649 = |\mathrm{N}(71 - 16\omega)|$.

$$71 - 16\omega = (24 + 20\omega)(5 - 3\omega) + (11 + 16\omega),$$

where $|\mathrm{N}(11 + 16\omega)| = 41 < 656 = |\mathrm{N}(24 + 20\omega)|$.

$$24 + 20\omega = (11 + 16\omega)(8 - 4\omega).$$

Consequently, $\delta = 11 + 16\omega$, and

$$11 + 16\omega = 71 - 16\omega - (24 + 20\omega)(5 - 3\omega)$$
$$= 71 - 16\omega - \left[95 + 4\omega - (71 - 16\omega) \right](5 - 3\omega)$$
$$= (-5 + 3\omega)(95 + 4\omega) + (6 - 3\omega)(71 - 16\omega).$$

Thus $\delta = 11 + 16\omega$, $\lambda = -5 + 3\omega$, and $\kappa = 6 - 3\omega$.

DEFINITION The integers α and β are said to be **relatively prime** if there exists a unit ϵ such that $\epsilon = GCD(\alpha, \beta)$.

If α and β are relatively prime, it follows from Theorem 2.18(ii) that $1 = GCD(\alpha, \beta)$.

DEFINITION If α and β are not zero, then $\delta \neq 0$ is a **least common multiple** of α and β if
(i) $\alpha \mid \delta$ and $\beta \mid \delta$;
(ii) whenever $\alpha \mid \gamma$ and $\beta \mid \gamma$, then $\delta \mid \gamma$.
We shall use the notation $\delta = LCM(\alpha, \beta)$ to denote that δ is a least common multiple of α and β.

THEOREM 2.19 *If α and β are not zero, then*
(i) *α and β have a least common multiple δ;*
(ii) *δ' is a least common multiple of α and β if and only if δ and δ' are associates;*
(iii) *$LCM(\alpha, \beta) = \alpha\beta/GCD(\alpha, \beta)$.*

We shall use the notation $lcm(a,b)$ in the sense of rational integer number theory; that is, $lcm(a,b)$ is the unique smallest positive rational integer divisible by both a and b. Since $lcm(a,b) = |ab|/gcd(a,b)$ and $gcd(a,b) = GCD(a,b)$, it is evident from Theorem 2.19(iii) that $lcm(a,b) = LCM(a,b)$.

EXAMPLE 2.5 From Theorem 2.19 and Example 2.4, we find that

$$LCM(95 + 4\omega, 71 - 16\omega) = (95 + 4\omega)(71 - 16\omega)/(11 + 16\omega)$$
$$= (6681 - 1300\omega)/(11 + 16\omega)$$
$$= (6681 - 1300\omega)(11 + 16\overline{\omega})/41$$
$$= (201187 - 121196\omega)/41$$
$$= 4907 - 2956\omega.$$

THEOREM 2.20 (Euclid's lemma)   *If $\alpha \mid \beta\gamma$ and GCD$(\alpha, \beta) = 1$, then $\alpha \mid \gamma$.*

DEFINITION   If π is not zero or a unit, then π is said to be a **prime** (of $\hat{Z}$) if whenever $\pi = \alpha\beta$, then one of α or β is a unit.

THEOREM 2.21   *If π is a prime and $\pi \mid \alpha_1\alpha_2 \cdots \alpha_n$, then π divides at least one of the α_i.*

THEOREM 2.22 (Unique factorization)   *If α is not zero or a unit, unit, then the following two conditions hold:*
(i) *There exist a finite number of primes $\pi_1, \pi_2, \ldots, \pi_s$ such that $\alpha = \pi_1\pi_2 \cdots \pi_s$.*
(ii) *If $\alpha = \pi_1\pi_2 \cdots \pi_s = \pi_1'\pi_2' \cdots \pi_t'$, where the π_i and π_j' are primes, then $s = t$, each π_i is an associate of some π_j', and conversely each π_j' is an associate of some π_i.*

THEOREM 2.23   *The ideal $M = \mu\hat{Z}$ is a maximal ideal of $\hat{Z}$ if and only if μ is a prime.*

THEOREM 2.24   *If $\pi_1, \pi_2, \ldots, \pi_n$ are nonassociated primes, a_1, $a_2, \ldots, a_n, b_1, b_2, \ldots, b_n$ are nonnegative, ϵ_1, ϵ_2 are units of $\hat{Z}$, $\alpha = \epsilon_1\pi_1^{a_1}\pi_2^{a_2} \cdots \pi_n^{a_n}$, and $\beta = \epsilon_2\pi_1^{b_1}\pi_2^{b_2} \cdots \pi_n^{b_n}$, then*

$$\text{GCD}(\alpha, \beta) = \pi_1^{c_1}\pi_2^{c_2} \cdots \pi_n^{c_n},$$

and

$$\text{LCM}(\alpha, \beta) = \pi_1^{d_1}\pi_2^{d_2} \cdots \pi_n^{d_n},$$

where $c_i = \min(a_i, b_i)$ and $d_i = \max(a_i, b_i)$, $i = 1, 2, \ldots, n$.

III   Units and Primes

Introduction

In this chapter we are primarily concerned with identifying the units and primes of $\hat{Z}$ and obtaining a prime factorization of an arbitrary integer α. However, the material of this chapter admits an easy solution of the Diophantine equation $x^2 + xy - y^2 = M$, and so we will take the time to consider this application.

In this and the succeeding chapters, we usually denote rational primes by the lowercase Latin letters p and q and primes of $\hat{Z}$ by the lowercase Greek letter π.

Units

THEOREM 3.1   *The positive units of $\hat{Z}$ are given by ω^n. The units ω^n, $n > 0$, are greater than 1 and tend monotonically to ∞ as n tends* to ∞. *The units ω^n, $n < 0$, are less than 1 and tend monotonically to 0 as n tends to $-\infty$. The negative units of $\hat{Z}$ are given by $-\omega^n$. Thus there are an infinite number of positive units and an infinite number of negative units.*

PROOF: Since $N(\omega) = \omega\bar{\omega} = -1$, then ω is a unit. As the units are a multiplicative group, it follows that ω^n is a unit for any n. Since $\omega = \frac{1}{2}(1 + \sqrt{5}) \approx 1.62 > 1$, the monotonicity properties are evident. It remains to show that these are the only positive units. A major step in this direction is to show that there is no unit ϵ satisfying $1 < \epsilon < \omega$. So let us suppose that $\epsilon = c + d\omega$ is a unit satisfying $1 < \epsilon \leq \omega$. We shall establish that $\epsilon = \omega$, and the desired conclusion will follow. Since $1 < c + d\omega \leq \omega$ and $|N(c + d\omega)| = |(c + d\omega)(c + d\bar{\omega})| = 1$, then we must have $|c + d\bar{\omega}| < 1$. Hence, $-1 < -c - d\bar{\omega} < 1$. By adding the two inequalities, $1 < c + d\omega \leq \omega$ and $-1 < -c - d\bar{\omega} < 1$, we obtain $0 < d\sqrt{5} < 1 + \omega \approx 2.62$. From this inequality, we conclude that $d = 1$. But then $1 < c + d\omega \leq \omega$ with $d = 1$ implies that $-0.62 \approx 1 - \omega < c \leq 0$. Consequently, $c = 0$, and so $\epsilon = c + d\omega = \omega$. Thus there is no unit ϵ satisfying $1 < \epsilon < \omega$.

Now if $\epsilon > 0$ is a unit and $\epsilon \neq \omega^n$ for any n, then, from the monotonicity properties, we must have $\omega^s < \epsilon < \omega^{s+1}$ for some s. Thus $1 < \epsilon\omega^{-s} < \omega$, and as the units are a group, $\epsilon\omega^{-s}$ would be a unit between 1 and ω, contrary to what was proved in the previous paragraph. Hence, if $\epsilon > 0$ is a unit, then $\epsilon = \omega^n$ for some n.

The statements concerning the negative units follow from the observation that $-\epsilon$ is a unit if and only if ϵ is a unit. Q.E.D.

COROLLARY 3.2 *The integer ϵ is a unit if and only if $\epsilon = \pm\omega^n$ for some n.*

Units and the Fibonacci Numbers

A unit ϵ may be expressed either in the form $\epsilon = c + d\omega$ or $\epsilon = \pm\omega^n$, and it is important that we be able to convert from one form to the other. There is a series of relationships between the units of $\hat{Z}$ and the Fibonacci numbers which make this conversion an easy matter. The relationships we have in mind are set forth in Theorem 3.3. The reader is no doubt familiar with the Fibonacci numbers, and so the following definition is a mere formality.

DEFINITION The sequence $F_0, F_1, F_2, \ldots$ defined recursively by $F_0 = 0$, $F_1 = 1$, and $F_{n+1} = F_n + F_{n-1}$, $n > 0$, is called the **Fibonacci sequence.** F_n is called the **nth Fibonacci number.**

THEOREM 3.3 *If $n > 0$, then*
(i) $\omega^n = F_{n-1} + F_n\omega$,
(ii) $\bar{\omega}^n = F_{n+1} - F_n\omega$,
(iii) $\omega^{-n} = (-1)^n(F_{n+1} - F_n\omega)$.

PROOF: We use mathematical induction to prove (i). When $n = 1$ we have $\omega^1 = 0 + 1\omega = F_0 + F_1\omega$, so that the formula is true for $n = 1$. Now suppose that (i) is true for $n = k$; i.e., $\omega^k = F_{k-1} + F_k\omega$. Then $\omega^{k+1} = (F_{k-1} + F_k\omega)\omega = F_{k-1}\omega + F_k\omega^2 = F_{k-1}\omega + F_k(\omega + 1) = F_k + (F_{k-1} + F_k)\omega = F_k + F_{k+1}\omega$, which is the desired result for $n = k + 1$. As for (ii), we note from (i) that $\bar{\omega}^n = F_{n-1} + F_n\bar{\omega} = F_{n-1} + F_n(1 - \omega) = (F_{n-1} + F_n) - F_n\omega = F_{n+1} - F_n\omega$. Finally, $\omega^{-1} = -\bar{\omega}$, so that $\omega^{-n} = (-1)^n\bar{\omega}^n = (-1)^n(F_{n+1} - F_n\omega)$. Q.E.D.

EXAMPLE 3.1 $\omega^5 = F_4 + F_5\omega = 3 + 5\omega$; $\omega^{-4} = (-1)^4(F_5 - F_4\omega) = 5 - 3\omega$.

When applying Theorem 3.3, it is expedient to have a short list of Fibonacci numbers. We therefore list in Appendix A the Fibonacci numbers F_1 through F_{40}. More extensive tables are available: Brosseau [2] lists F_1 through F_{100} and gives their rational prime decompositions; Jardan [12] lists F_1 through F_{385} and gives the rational prime decompositions of F_1 through F_{171}. We will return to the subject of Fibonacci numbers in Chapter 9.

Primes

THEOREM 3.4 *If q is a rational prime and $|N(\alpha)| = q$, then α is a prime.*

PROOF: Suppose that $\alpha = \beta\gamma$. Then $q = |N(\alpha)| = |N(\beta)| \cdot |N(\gamma)|$, and since q is a rational prime, we must have $|N(\beta)| = 1$ or $|N(\gamma)| = 1$. Thus either β or γ is a unit, and we conclude that α is a prime. Q.E.D.

THEOREM 3.5 *If π is a prime, then there exists one and only one rational prime p such that $\pi \mid p$.*

PROOF: Since π divides $|N(\pi)|$, there exist positive rational integers divisible by π. Let p be the smallest such. If p is not a rational prime, then $p = n_1 n_2$, where $1 < n_1 \leq n_2 < p$. Thus we would have $\pi \mid n_1 n_2$, and so $\pi \mid n_1$ or $\pi \mid n_2$, by Theorem 2.21. Since the last statement contradicts the minimality of p, we are forced to conclude that p is a prime.

Now suppose, to the contrary, π divides two distinct rational primes p and q. Since $\gcd(p, q) = 1$, then $1 = px + qy$ for some x and y. This would entail that $\pi \mid 1$ because $\pi \mid (px + qy)$, and so π would be a unit. But π is not a unit, and so π divides precisely one rational prime. Q.E.D.

The next four lemmas establish the additional preliminaries needed for the classification of the primes given in Theorem 3.10.

LEMMA 3.6 *The integer $2 + \omega$ is a prime; and if a prime π divides 5, then π is an associate of $2 + \omega$.*

PROOF: Since $N(2 + \omega) = 2^2 + 2 - 1 = 5$, then $2 + \omega$ is a prime, by Theorem 3.4. Also, $5\omega^2 = 5(1 + \omega) = (2 + \omega)^2$, and so $5 = \omega^{-2}(2 + \omega)^2$. Thus, if a prime π divides 5, then $\pi \mid (2 + \omega)$, by Theorem 2.21 (π cannot divide ω^{-1}). Consequently, $\pi = \epsilon(2 + \omega)$; and ϵ is a unit, for otherwise, π would not be a prime. Accordingly, π is an associate of $2 + \omega$. Q.E.D.

From Lemma 3.6 we observe, in particular, that $\sqrt{5}$ is an associate of $2 + \omega$. An easy calculation yields $\sqrt{5} = \omega^{-1}(2 + \omega)$.

LEMMA 3.7 *If the rational prime q is not a prime, then $N(\pi) = q$ for some prime π.*

PROOF: Since q is not a prime, there exist nonunits α and β such that $q = \alpha\beta$. Then $q^2 = N(\alpha)N(\beta)$, $|N(\alpha)| > 1$, $|N(\beta)| > 1$, and $N(\alpha)$ and $N(\beta)$ are rational integers. Hence, since q is a rational prime, then $N(\alpha) = \pm q$. Set $\pi = \alpha$ or $\pi = \omega\alpha$ according as $N(\alpha) = q$ or $N(\alpha) = -q$. Then $N(\pi) = q$ and, by Theorem 3.4, π is a prime. Q.E.D.

LEMMA 3.8 *If p is a rational prime and $p \equiv \pm 2$ (mod 5), then p is a prime. Also, any prime π that divides p is an associate of p.*

PROOF: Suppose, contrariwise, that p is not a prime. Then, by Lemma 3.7, $N(\pi) = p$ for some prime $\pi = c + d\omega$. Thus $4p = 4N(\pi) = (2c + d)^2 - 5d^2$, and so $(2c + d)^2 \equiv 4p \equiv -p \equiv \pm 2$ (mod 5). But $(2c + d)^2 \equiv \pm 2$ (mod 5) is impossible, since 2 and -2 are not quadratic residues modulo 5. Hence our contrary assumption is false, and p is a prime. The second assertion of the lemma is clear. Q.E.D.

LEMMA 3.9 *If q is a rational prime and $q \equiv \pm 1$ (mod 5), then $N(\pi) = q$ for some prime π. Moreover, if π is any integer such that $N(\pi) = q$, then*
(i) *π and $\bar{\pi}$ are primes;*
(ii) *π and $\bar{\pi}$ are not associates;*
(iii) *if a prime π_1 divides q, then π_1 is an associate of π or $\bar{\pi}$.*

PROOF: The first statement of the lemma will follow from Lemma 3.7 provided we show that q is not a prime. Since -1 and 1 are quadratic residues modulo 5, then $(q|5) = (\pm 1|5) = 1$ ($(q|5)$ is the Legendre symbol). Thus, by quadratic reciprocity, $(5|q) = 1$. Consequently, there exists a b such that $b^2 \equiv 5$ (mod q), and so $q|(b + \sqrt{5})(b - \sqrt{5})$. Now, if q is a prime, we must have $q|(b + \sqrt{5})$ or $q|(b - \sqrt{5})$, by Theorem 2.21. Thus either $(b + \sqrt{5})/q$ or $(b - \sqrt{5})/q$ is an integer. But no integer is of either of these two forms. Hence we conclude that q is not a prime.

Assertion (i) follows from Theorem 3.4, since $N(\pi) = N(\bar{\pi}) = q$. Assertion (iii) is also clear. For if $\pi_1 | q$, then $\pi_1 | \pi\bar{\pi}$; and so $\pi_1 | \pi$ or $\pi_1 | \bar{\pi}$. In the former case π_1 is an associate of π, and in the latter case π_1 is an associate of $\bar{\pi}$. Only assertion (ii) remains to be proved. We argue by contradiction. Suppose that $\pi = c + d\omega$ and $\bar{\pi} = c + d\bar{\omega}$ are associates. Then $\pi/\bar{\pi}$ is a unit, and, in particular, $\pi/\bar{\pi}$ is an integer. Now

$$\frac{\pi}{\bar{\pi}} = \frac{\pi^2}{\pi\bar{\pi}} = \frac{\pi^2}{q} = \frac{c^2 + d^2}{q} + \frac{2cd + d^2}{q}\omega$$

is an integer if and only if $c^2 + d^2 \equiv 0$ (mod q) and $2cd + d^2 \equiv 0$ (mod q). From these two congruences we obtain

$$c^2 \equiv -d^2 \equiv 2cd \;(\text{mod } q). \tag{3.1}$$

Also, we have $c^2 + cd - d^2 = N(\pi) = q \equiv 0$ (mod q), and so it follows from (3.1) that $5cd \equiv 0$ (mod q), and hence $cd \equiv 0$ (mod q). Consequently, by (3.1), we have $c \equiv d \equiv 0$ (mod q). But then q^2 divides $N(\pi) = c^2 + cd - d^2$, which is impossible since $N(\pi) = q$. This contradiction shows that π and $\bar{\pi}$ are not associates. Q.E.D.

THEOREM 3.10 (Classification of primes) *If q is a rational prime and $q \equiv \pm 1$ (mod 5), then let $N(\pi) = q$. (The existence of π follows from Lemma 3.9.) Then the primes of $\hat{Z}$ are*
(i) *$2 + \omega$ and its associates;*
(ii) *the rational primes p, $p \equiv \pm 2$ (mod 5), and their associates;*
(iii) *the nonassociated factors π and $\bar{\pi}$ of the rational primes q, $q \equiv \pm 1$ (mod 5), and their associates.*
PROOF: Any rational prime except 5 is congruent modulo 5 to one of the rational integers -1, 1, -2, or 2. Thus the conclusions of the theorem follow immediately from Theorem 3.5 and Lemmas 3.6, 3.8, and 3.9. Q.E.D.

COROLLARY 3.11 *The number 5 is the only rational prime which is divisible by the square of a prime.*
PROOF: We have already seen that $5 = \omega^{-2}(2 + \omega)^2$, so that 5 is divisible by the square of a prime. If p is a rational prime and $p \equiv \pm 2$ (mod 5), then p is a prime and hence is not divisible by the square of a prime. If q is a rational prime of the form $q \equiv \pm 1$ (mod 5), then $N(\pi) = q$ for some prime π. Thus, if π_1 is a prime and $\pi_1^2 \mid q$, then $\pi_1^2 \mid \pi\bar{\pi}$; and so $\pi_1 \mid \pi$ and $\pi_1 \mid \bar{\pi}$. This would mean that π and $\bar{\pi}$ are associates, which is impossible, by Lemma 3.9. Thus, $\pi_1^2 \nmid q$, and the proof is complete. Q.E.D.

In Appendix B a prime π_p such that $\pi_p \mid p$ is given for each rational prime $p < 32,771$. When p is of the form $p \equiv \pm 2 \pmod 5$, then $\pi_p = p$; when p is not of this form, then the π_p given satisfies $N(\pi_p) = p$. The following example illustrates how the primes in Appendix B were determined when $p \equiv \pm 1 \pmod 5$.

EXAMPLE 3.2 The rational prime 881 is congruent to 1 modulo 5, and hence we can find and integer $\frac{1}{2}a + \frac{1}{2}b\sqrt{5}$ with norm 881. That is, we can find nonzero a and b with the same parity such that

$$a^2 - 5b^2 = 4(881) = 3524. \tag{3.2}$$

We shall, in fact, determine the smallest positive pair (a, b) satisfying (3.2). It is evident that $|a| \geq \sqrt{3524} \approx 59.4$, and the accompanying table is self-explanatory.

a	b	$a^2 - 5b^2$
60	2	3580
60	4	3520
61	3	3676
61	5	3596
61	7	3476
62	6	3664
62	8	3524

Thus we conclude from the table that the integer $27 + 8\omega = \frac{1}{2}(62) + \frac{1}{2}(8)\sqrt{5}$ has norm 881, and so $27 + 8\omega$ and $27 + 8\overline{\omega}$ are two nonassociated primes that divide the rational prime 881.

$N(c + d\omega) = M$ *and the Factorization of* $c + d\omega$

Now that the units and primes have been cataloged we consider the determination of a prime factorization of a nonzero integer

$c + d\omega$. The following theorem shows that a prime factorization of $c + d\omega$ is closely related to the rational integer prime factorization of $\mathrm{N}(c + d\omega)$.

THEOREM 3.12 *If $c + d\omega$ is a nonzero integer and $\mathrm{N}(c + d\omega)$ $= M$, then M is of the form*

$$M = (\pm 1)5^a p_1^{2b_1} p_2^{2b_2} \cdots p_r^{2b_r} q_1^{c_1} q_2^{c_2} \cdots q_s^{c_s}, \qquad (3.3)$$

where the p_i are distinct rational primes of the form $p_i \equiv \pm 2 \pmod 5$, the q_i are distinct rational primes of the form $q_i \equiv \pm 1 \pmod 5$, and all exponents are nonnegative. Conversely, if M is of the form (3.3), then there exists a nonzero integer $c + d\omega$ such that $\mathrm{N}(c + d\omega)$ $= M$; and $c + d\omega$ is of the form

$$c + d\omega = \epsilon(2 + \omega)^a p_1^{b_1} p_2^{b_2} \cdots p_r^{b_r} \left(\pi_1^{j_1} \bar{\pi}_1^{c_1 - j_1} \right) \cdot \qquad (3.4)$$

$$\left(\pi_2^{j_2} \bar{\pi}_2^{c_2 - j_2} \right) \cdots \left(\pi_s^{j_s} \bar{\pi}_s^{c_s - j_s} \right),$$

where ϵ is a unit, π_i is a prime such that $\mathrm{N}(\pi_i) = q_i$, and $j_i \in \{0, 1, 2, \ldots, c_i\}$. If M is positive, then $\mathrm{N}(c + d\omega) = M$ if and only if $c + d\omega$ is of the form (3.4) and $\epsilon = \pm \omega^{2n}$. If M is negative, then $\mathrm{N}(c + d\omega) = M$ if and only if $c + d\omega$ is of the form (3.4) and $\epsilon = \pm \omega^{2n+1}$.

PROOF: All assertions in the theorem are immediate consequences of the unique factorization property, the basic properties of norms and units, and the classification of the primes given by Theorem 3.10. Q.E.D.

Two corollaries of Theorem 3.12 are worthy of mention.

COROLLARY 3.13 *If $\mathrm{N}(x + y\omega) = M$, where M is given by (3.3), then there are exactly $(c_1 + 1)(c_2 + 1) \cdots (c_s + 1)$ nonassociated solutions $c + d\omega$ of $\mathrm{N}(x + y\omega) = M$.*
PROOF: Each possible s-tuple $(j_1, j_2, \ldots, j_s)$ yields a nonassociated solution, and these $(c_1 + 1)(c_2 + 1) \cdots (c_s + 1)$ possible s-tuples exhaust the nonassociated solutions $c + d\omega$. Q.E.D.

COROLLARY 3.14 *If $c + d\omega \neq 0$ and $\gcd(c, d) = 1$, then there exist distinct rational primes $q_1, q_2, \ldots, q_s$ of the form $q_i \equiv \pm 1$ (mod 5), primes $\pi_1, \pi_2, \ldots, \pi_s$ such that $\mathrm{N}(\pi_i) = q_i$, a unit ϵ, and nonnegative exponents $a, c_1, c_2, \ldots, c_s$ such that*

$$c + d\omega = \epsilon(2 + \omega)^a \pi_1^{c_1} \pi_2^{c_2} \cdots \pi_s^{c_s}. \qquad (3.5)$$

PROOF: Since $\gcd(c,d) = 1$, then in (3.4) we have $b_1 = b_2 = \ldots = b_r = 0$. For otherwise, some p_i would divide $c + d\omega$, and so c and d would have the common factor p_i, contrary to the hypothesis that $\gcd(c,d) = 1$. Similarly, either $j_i = 0$ or $j_i = c_i$ for each i, $i = 1$, $2, \ldots, s$. (For otherwise, c and d would have a common factor q_i.) The desired result now follows directly from (3.4), upon noting that each π_i in (3.5) is either the π_i or the $\bar{\pi}_i$ of (3.4). Q.E.D.

EXAMPLE 3.3 We obtain a prime factorization of $-1152 + 1324\omega$ in terms of the primes of Appendix B and the unit ω. We first note that $-1152 + 1324\omega = 2^2(-288 + 331\omega)$, $\gcd(-288, 331) = 1$, and $N(-288 + 331\omega) = -121945 = -5 \cdot 29^3$. Thus it follows from Theorem 3.12, Corollary 3.14, and Appendix B that $-288 + 331\omega = \epsilon(2 + \omega)\pi^3$, where ϵ is a unit, and $\pi = 5 + \omega$ or $\pi = 5 + \bar{\omega}$. A computation shows that $5 + \omega$ divides $-288 + 331\omega$, and so $-288 + 331\omega = \epsilon(2 + \omega)(5 + \omega)^3$. Another computation yields

$$\epsilon = \frac{-288 + 331\omega}{(2 + \omega)(5 + \omega)^3} = -3 + 2\omega,$$

and since $\omega^{-3} = (-1)^3(F_4 - F_3\omega) = -(3 - 2\omega) = -3 + 2\omega = \epsilon$, we obtain the prime factorization

$$-1152 + 1324\omega = \omega^{-3}(2 + \omega)2^2(5 + \omega)^3.$$

The Equation $x^2 + xy - y^2 = M$

THEOREM 3.15 *Let $M \neq 0$ be a rational integer. The Diophantine equation $x^2 + xy - y^2 = M$ has a solution (c,d) in rational integers if and only if M is of the form (3.3) of Theorem 3.12. Moreover, (c,d) is a rational integer solution if and only if $c + d\omega$ is a solution of $N(x + y\omega) = M$.*

PROOF: Immediate from Theorem 3.12, since $N(x + y\omega) = x^2 + xy - y^2$. Q.E.D.

EXAMPLE 3.4 We find all rational integer solutions (c, d) of $x^2 + xy - y^2 = 45980$. Since $45980 = 5 \cdot 2^2 \cdot 11^2 \cdot 19$, it follows that $c + d\omega$ is a solution of $N(x + y\omega) = 45980$ if and only if

$$c + d\omega = (\pm\omega^{2n})(2 + \omega)(2)(3 + \omega)^i(3 + \overline{\omega})^{2-i} \cdot$$
$$(4 + \omega)^j(4 + \overline{\omega})^{1-j},$$

where $i \in \{0, 1, 2\}$, and $j \in \{0, 1\}$. Thus the six nonassociated solutions of $N(x + y\omega) = 45980$ are given by

$$x_1 + y_1\omega = (2 + \omega)(2)(3 + \overline{\omega})^2(4 + \overline{\omega}) = 278 - 86\omega,$$

$$x_2 + y_2\omega = (2 + \omega)(2)(3 + \overline{\omega})^2(4 + \omega) = 208 + 14\omega,$$

$$x_3 + y_3\omega = (2 + \omega)(2)(3 + \omega)(3 + \overline{\omega})(4 + \overline{\omega}) = 198 + 44\omega,$$

$$x_4 + y_4\omega = (2 + \omega)(2)(3 + \omega)(3 + \overline{\omega})(4 + \omega) = 198 + 154\omega,$$

$$x_5 + y_5\omega = (2 + \omega)(2)(3 + \omega)^2(4 + \overline{\omega}) = 208 + 194\omega,$$

$$x_6 + y_6\omega = (2 + \omega)(2)(3 + \omega)^2(4 + \omega) = 278 + 364\omega.$$

Thus $(278, -86)$, $(208, 14)$, $(198, 44)$, $(198, 154)$, $(208, 194)$, and $(278, 364)$ are solutions of $x^2 + xy - y^2 = 45980$. Any and all other solutions are given by (c, d), where $c + d\omega = (\pm\omega^{2n})(x_k + y_k\omega)$, $k = 1, 2, \ldots, 6$.

Number of Divisors of $c + d\omega$

We conclude the present chapter with the standard result on the number of divisors of a nonzero integer.

NOTATION If $\beta \neq 0$, then $\hat{\tau}(\beta)$ denotes the number of nonassociated divisors of β.

We will use the notation $\tau(a)$ in the sense of rational integer theory; that is, $\tau(a)$ denotes the number of positive rational integers that divide a.

THEOREM 3.16

(i) *If ϵ is a unit, then $\hat{\tau}(\epsilon) = 1$.*

(ii) *If $\beta \neq 0$ is a nonunit and $\beta = \epsilon \pi_1^{e_1} \pi_2^{e_2} \cdots \pi_r^{e_r}$, where the π_i are nonassociated primes, ϵ is a unit, and the e_i are positive, then $\hat{\tau}(\beta) = (e_1 + 1)(e_2 + 1) \cdots (e_r + 1)$.*

PROOF: Assertion (i) is trivial. Now $\alpha \mid \beta$ if and only if $\alpha = \epsilon_1 \pi_1^{f_1} \pi_2^{f_2} \cdots \pi_r^{f_r}$, where ϵ_1 is a unit and $0 \leq f_i \leq e_i$. Also, each different r-tuple $(f_1, f_2, \ldots, f_r)$ yields a different nonassociated divisor of β. Since there are $(e_1 + 1)(e_2 + 1) \cdots (e_r + 1)$ possible r-tuples, the desired result follows. Q.E.D.

EXAMPLE 3.5 Since $-1152 + 1324\omega = \omega^{-3}(2 + \omega)(2^2)(5 + \omega)^3$ from Example 3.3, then $\hat{\tau}(-1152 + 1324\omega) = (2)(3)(4) = 24$.

IV Discriminants and Integral Bases

Introduction

According to Theorem 2.12, every ideal I of $\hat{Z}$ is principal; that is, if I is an ideal, then there exists a μ such that $I = \mu\hat{Z} = \{\mu\alpha : \alpha \in \hat{Z}\}$. In this chapter we consider another useful way of representing a nonzero ideal I; namely, representation of I by an integral basis.

If α and β are two elements of a nonzero ideal I, we say that $\{\alpha, \beta\}$ is an *integral basis* for I provided every γ in I can be uniquely expressed in the form $\gamma = a\alpha + b\beta$ (a and b rational integers). If $\{\alpha, \beta\}$ is an integral basis for I, then it is plain that $I = \{a\alpha + b\beta : a, b \in Z\}$. As we will see in Corollary 4.4, every nonzero ideal has an integral basis. As a matter of fact, the ideal $\mu\hat{Z}$, $\mu \neq 0$, possesses an integral basis of the form $\{m, s + n\omega\}$, where $m > 0$ and $n > 0$. An integral basis of this form is called a *canonical basis* for $\mu\hat{Z}$, and, as will be shown in Theorem 4.10, m and n are uniquely determined by μ.

The *discriminant*, $\Delta(\alpha, \beta)$, of α and β is defined by $\Delta(\alpha, \beta) = (\alpha\bar{\beta} - \beta\bar{\alpha})^2$. We have the following important property (Theorem 4.7): If $\mu \neq 0$ and $\{\alpha, \beta\}$ is an integral basis for $\mu\hat{Z}$, then $\Delta(\alpha, \beta) = 5(N(\mu))^2$. Calling $\Delta(\alpha, \beta)$ the *discriminant* of the integral basis

$\{\alpha, \beta\}$, we see that any two integral bases for $\mu\hat{Z}$ have the same discriminant $5(N(\mu))^2$. This result allows us to conclude that $|N(\mu)|$ = mn when $\{m, s + n\omega\}$ is a canonical basis for $\mu\hat{Z}$. This and other results on canonical bases considered in this chapter will be of considerable help in Chapter 5 when we determine complete residue systems for various moduli.

The material of this chapter is a synthesis of results found in Pollard and Diamond [17], Reid [18], and Jones [13], pp. 153–158. In particular, the material on canonical bases is heavily influenced by Jones.

Discriminants

DEFINITION The **discriminant** of the two integers α and β, denoted by $\Delta(\alpha, \beta)$, is given by $\Delta(\alpha, \beta) = (\alpha\bar{\beta} - \beta\bar{\alpha})^2$.

In Theorem 4.1 we show that $\Delta(\alpha, \beta) = \Delta(\beta, \alpha)$, so that the above definition is in no way ambiguous. It should be noted that if

$$C = \begin{pmatrix} \alpha & \bar{\alpha} \\ \beta & \bar{\beta} \end{pmatrix},$$

then $\Delta(\alpha, \beta) = (\det C)^2$.

THEOREM 4.1
(i) $\Delta(\alpha, \beta) = \Delta(\beta, \alpha)$.
(ii) $\Delta(\alpha, \beta)$ *is a nonnegative rational integer.*
(iii) $\{\alpha, \beta\}$ *is a basis for* $Q(\sqrt{5})$ *over* Q *if and only if* $\Delta(\alpha, \beta) \neq 0$.
PROOF: Part (i) follows from $\Delta(\alpha, \beta) = (\alpha\bar{\beta} - \beta\bar{\alpha})^2 = (\beta\bar{\alpha} - \alpha\bar{\beta})^2 = \Delta(\beta, \alpha)$. It is evident that $\Delta(\alpha, \beta)$ is an integer, and since $\Delta(\alpha, \beta)$ is the square of a real number, then $\Delta(\alpha, \beta) \geq 0$. Thus, to establish (ii), it suffices to verify that $\Delta(\alpha, \beta)$ is a rational number, because the only integers of $\hat{Z}$ that are rational numbers are the rational integers. According to Theorem 2.5(ii), it is enough to show that $\Delta(\alpha, \beta) = \overline{\Delta(\alpha, \beta)}$; and this follows from

$$\overline{\Delta(\alpha, \beta)} = \overline{\left(\alpha\overline{\beta} - \beta\overline{\alpha}\right)^2} = \left(\overline{\alpha}\beta - \overline{\beta}\alpha\right)^2$$

$$= \left(\alpha\overline{\beta} - \beta\overline{\alpha}\right)^2 = \Delta(\alpha, \beta).$$

In order to establish (iii), we prove the contrapositive: $\{\alpha, \beta\}$ is not a basis for $Q(\sqrt{5})$ over Q if and only if $\Delta(\alpha, \beta) = 0$. In the first place, if α or β is zero, then $\{\alpha, \beta\}$ is not a basis and $\Delta(\alpha, \beta) = 0$. So we may assume in the proof that neither α nor β is zero. If $\Delta(\alpha, \beta) = 0$, then $\alpha\overline{\beta} - \beta\overline{\alpha} = 0$; and so $\alpha/\beta = \overline{(\alpha/\beta)}$. Thus, by Theorem 2.5(ii), $\alpha/\beta = c$, where c is a rational number. But then $1\alpha - c\beta = 0$, which shows that α and β are linearly dependent over Q. Accordingly, if $\Delta(\alpha, \beta) = 0$, then $\{\alpha, \beta\}$ is not a basis. Conversely, if $\Delta(\alpha, \beta)$ is not a basis, then $\alpha = c\beta$ for some rational number c. Then $\Delta(\alpha, \beta) = (\alpha\overline{\beta} - \beta\overline{\alpha})^2 = (c\beta\overline{\beta} - \beta c\overline{\beta})^2 = 0$, and the proof is complete. Q.E.D.

THEOREM 4.2 *If*

$$A = \begin{pmatrix} a & b \\ c & d \end{pmatrix} \quad \text{and} \quad \begin{pmatrix} \alpha \\ \beta \end{pmatrix} = A\begin{pmatrix} \gamma \\ \delta \end{pmatrix},$$

then $\Delta(\alpha, \beta) = (\det A)^2\Delta(\gamma, \delta)$.

PROOF: If we set

$$C = \begin{pmatrix} \alpha & \overline{\alpha} \\ \beta & \overline{\beta} \end{pmatrix} \quad \text{and} \quad D = \begin{pmatrix} \gamma & \overline{\gamma} \\ \delta & \overline{\delta} \end{pmatrix},$$

then we have $C = AD$; so that $\det C = (\det A)(\det D)$. Hence $\Delta(\alpha, \beta) = (\det C)^2 = (\det A)^2(\det D)^2 = (\det A)^2\Delta(\gamma, \delta)$. Q.E.D.

Integral Bases

DEFINITION If α and β are elements of an ideal I of $\hat{Z}$, then $\{\alpha, \beta\}$ is said to be an **integral basis** for I if every γ in I can be uniquely represented in the form $\gamma = c\alpha + d\beta$ (c and d rational integers).

The ideal $\{0\}$ does not have an integral basis. As we shall prove in Corollary 4.4, it is the only ideal of $\hat{Z}$ which does not. It is evident from the definition of an ideal that $I = \{c\alpha + d\beta : c, d \in Z\}$ whenever $\{\alpha, \beta\}$ is an integral basis for I.

NOTATION If $\{\alpha, \beta\}$ is an integral basis for the ideal I, then we define $\langle \alpha, \beta \rangle$ to be $\langle \alpha, \beta \rangle = \{c\alpha + d\beta : c, d \in Z\}$.

Thus, if I is an ideal of $\hat{Z}$, then the statement $I = \langle \alpha, \beta \rangle$ implies that
(i) α and β are in I;
(ii) if $\gamma \in I$, there exist unique rational integers c and d such that $\gamma = c\alpha + d\beta$;
(iii) $I = \{c\alpha + d\beta : c, d \in Z\}$.

THEOREM 4.3 If $\mu \neq 0$, then $\mu\hat{Z} = \langle \mu, \mu\omega \rangle$ and $\Delta(\mu, \mu\omega) = 5(\mathrm{N}(\mu))^2$.
PROOF: In the first place, we have $\Delta(\mu, \mu\omega) = (\mu\bar{\mu}\bar{\omega} - \mu\omega\bar{\mu})^2 = (\bar{\omega} - \omega)^2(\mu\bar{\mu})^2 = 5(\mathrm{N}(\mu))^2$. This establishes the second assertion of the theorem. In order to complete the proof, we must show that $\{\mu, \mu\omega\}$ is an integral basis for the ideal $\mu\hat{Z}$. Since μ and $\mu\omega$ are in $\mu\hat{Z}$, it suffices to show that any γ in $\mu\hat{Z}$ can be uniquely represented in the form $\gamma = c\mu + d\mu\omega$. But if γ is in $\mu\hat{Z}$, there exists a $c + d\omega$ such that $\gamma = \mu(c + d\omega) = c\mu + d(\mu\omega)$; and since $\Delta(\mu, \mu\omega) = 5(\mathrm{N}(\mu))^2 \neq 0$, it follows from Theorem 4.1(iii) that c and d are unique. Q.E.D.

COROLLARY 4.4 If I is a nonzero ideal of $\hat{Z}$, then I has an integral basis.
PROOF: All ideals in $\hat{Z}$ are principal ideals, and thus $I = \mu\hat{Z}$ for some $\mu \neq 0$. Therefore, by Theorem 4.3, $\{\mu, \mu\omega\}$ is an integral basis for I. Q.E.D.

THEOREM 4.5 If I is a nonzero ideal and $I = \langle \alpha, \beta \rangle$, then α and β are linearly independent over Q, and $\Delta(\alpha, \beta)$ is a positive rational integer.
PROOF: In view of Theorem 4.1, it is enough to show that α and β are linearly independent over Q. Therefore suppose that $r\alpha + s\beta = 0$ for some r and s in Q. There exist rational integers c_1, c_2, d_1, d_2,

$c_2 d_2 \neq 0$, such that $r = c_1/c_2$ and $s = d_1/d_2$. Upon clearing $r\alpha + s\beta = 0$ of fractions, we obtain $c_1 d_2 \alpha + c_2 d_1 \beta = 0$. Thus, since 0 is in I, it follows from the uniqueness of the representation of 0 that $c_1 = d_1 = 0$. Accordingly, $r = s = 0$ and the linear independence of α and β follows. Q.E.D.

THEOREM 4.6 *Let I be a nonzero ideal and $I = \langle \gamma, \delta \rangle$. Then $I = \langle \alpha, \beta \rangle$ if and only if there exists a matrix*

$$A = \begin{pmatrix} a & b \\ c & d \end{pmatrix}$$

such that $\det A = \pm 1$ *and*

$$\begin{pmatrix} \alpha \\ \beta \end{pmatrix} = A \begin{pmatrix} \gamma \\ \delta \end{pmatrix}.$$

Moreover, if $\langle \alpha, \beta \rangle = \langle \gamma, \delta \rangle$, *then* $\Delta(\alpha, \beta) = \Delta(\gamma, \delta)$.
PROOF: First suppose that $I = \langle \alpha, \beta \rangle$. Then α and β are in I, and since $I = \langle \gamma, \delta \rangle$, there exist a, b, c, and d such that $\alpha = a\gamma + b\delta$ and $\beta = c\gamma + d\delta$. Equivalently,

$$\begin{pmatrix} \alpha \\ \beta \end{pmatrix} = A \begin{pmatrix} \gamma \\ \delta \end{pmatrix}, \text{ where } A = \begin{pmatrix} a & b \\ c & d \end{pmatrix}.$$

Reversing the roles of $\langle \alpha, \beta \rangle$ and $\langle \gamma, \delta \rangle$, we have

$$\begin{pmatrix} \gamma \\ \delta \end{pmatrix} = B \begin{pmatrix} \alpha \\ \beta \end{pmatrix}, \text{ where } B = \begin{pmatrix} e & f \\ g & h \end{pmatrix}.$$

Consequently,

$$\begin{pmatrix} \alpha \\ \beta \end{pmatrix} = AB \begin{pmatrix} \alpha \\ \beta \end{pmatrix}.$$

Moreover, since $I = \langle \alpha, \beta \rangle$, then $\alpha = 1 \cdot \alpha + 0 \cdot \beta$ and $\beta = 0 \cdot \alpha + 1 \cdot \beta$ is the only way that α and β can be expressed as a linear combination of α and β using rational integer coefficients. Thus, if

$$\begin{pmatrix} \alpha \\ \beta \end{pmatrix} = C \begin{pmatrix} \alpha \\ \beta \end{pmatrix}.$$

where C has rational integer entries, then

$$C = \begin{pmatrix} 1 & 0 \\ 0 & 1 \end{pmatrix}.$$

Consequently,

$$\begin{pmatrix} 1 & 0 \\ 0 & 1 \end{pmatrix} = AB,$$

and so $(\det A)(\det B) = 1$. Since both determinants have rational integer values, then $\det A = \pm 1$.

Next suppose that the matrix A satisfies the given conditions. Since I is an ideal, then $\alpha = a\gamma + b\delta$ and $\beta = c\gamma + d\delta$ belong to I. Also, if η is in I, then there exist x and y such that

$$\eta = x\gamma + y\delta = (x, y)\begin{pmatrix} \gamma \\ \delta \end{pmatrix}.$$

Now A^{-1} exists and moreover it has rational integer entries, because $\det A = \pm 1$. Thus

$$\begin{pmatrix} \gamma \\ \delta \end{pmatrix} = A^{-1}\begin{pmatrix} \alpha \\ \beta \end{pmatrix}, \quad \text{and so } \eta = (x, y)A^{-1}\begin{pmatrix} \alpha \\ \beta \end{pmatrix} = x'\alpha + y'\beta,$$

where x' and y' are rational integers. By Theorems 4.2 and 4.5, $\Delta(\alpha, \beta) = (\det A)^2\Delta(\gamma, \delta) = \Delta(\gamma, \delta) \neq 0$; and thus, by Theorem 4.1, α and β are linearly independent over Q. Consequently, the representation $\eta = x'\alpha + y'\beta$ is unique, and we therefore conclude that $I = \langle \alpha, \beta \rangle$.

The last statement of the theorem follows from Theorem 4.2, since $(\det A)^2 = 1$. Q.E.D.

THEOREM 4.7 *If* $\mu \neq 0$ *and* $\mu\hat{Z} = \langle \alpha, \beta \rangle$, *then* $\Delta(\alpha, \beta) = 5(\mathrm{N}(\mu))^2$.

PROOF: It follows from Theorems 4.3 and 4.6 that $\Delta(\alpha, \beta) = \Delta(\mu, \mu\omega) = 5(\mathrm{N}(\mu))^2$. Q.E.D.

The last theorem motivates the following definition:

DEFINITION If $\mu \neq 0$, then $5(\mathrm{N}(\mu))^2$ is called the **discriminant** of the ideal $\mu\hat{Z}$.

Canonical Basis for $\mu\hat{Z}$

The major results of this section are Theorems 4.10 and 4.11. These two theorems will be needed in Chapter 5 when we consider

complete residue systems. The proof of Theorem 4.10 will be based
on the two lemmas which follow. The ideal of Z generated by c
will be denoted by cZ; that is, $cZ = \{cz : z \in Z\}$.

LEMMA 4.8 *If $\mu \neq 0$, m is the smallest positive rational integer in
$\mu\hat{Z}$, and n is the smallest positive rational integer in $\{d : c + d\omega
\in \mu\hat{Z}\}$, then $mZ = Z \cap \mu\hat{Z}$, and $nZ = \{d : c + d\omega \in \mu\hat{Z}\}$.*
PROOF: Since $|N(\mu)| > 0$ is in $\mu\hat{Z}$ and $|N(\mu)|\omega$ is in $\{d : c + d\omega
\in \mu\hat{Z}\}$, then $\mu\hat{Z}$ and $\{d : c + d\omega \in \mu\hat{Z}\}$ contain positive rational
integers. Thus the m and n of the lemma exist. Using the fact that
$\mu\hat{Z}$ is an ideal of $\hat{Z}$, it is trivial to verify that $Z \cap \mu\hat{Z}$ and
$\{d : c + d\omega \in \mu\hat{Z}\}$ are ideals of Z. The conclusion of the lemma
now follows since a nonzero ideal of Z is generated by the smallest
positive element in the ideal. Q.E.D.

LEMMA 4.9 *If $\mu \neq 0$ and $\mu\hat{Z} = \langle m, s + n\omega \rangle$, where $m > 0$ and
$n > 0$, then $|N(\mu)| = mn$, $mZ = Z \cap \mu\hat{Z}$, and $nZ = \{d : c + d\omega
\in \mu\hat{Z}\}$.*
PROOF: We have $\Delta(m, s + n\omega) = [m(s + n\bar{\omega}) - m(s + n\omega)]^2$
$= (\bar{\omega} - \omega)^2(mn)^2 = 5(mn)^2$. However, $\Delta(m, s + n\omega) = 5(N(\mu))^2$, by
Theorem 4.7, and so $|N(\mu)| = mn$. By the definition of integral
basis, z is in $Z \cap \mu\hat{Z}$ if and only if $z = am + 0 \cdot (s + n\omega) = am$ for
some a in Z. Accordingly, $mZ = Z \cap \mu\hat{Z}$. Similarly, $c + d\omega$ is in
$\mu\hat{Z}$ if and only if $c + d\omega = am + b(s + n\omega) = am + bs + bn\omega$ for
some a and b in Z. Thus d is in $\{d : c + d\omega \in \mu\hat{Z}\}$ if and only if d
is a multiple of n; and so $nZ = \{d : c + d\omega \in \mu\hat{Z}\}$. Q.E.D.

THEOREM 4.10 *If $\mu \neq 0$, then $\mu\hat{Z}$ has an integral basis $\{m, s +
n\omega\}$, where $m > 0$ and $n > 0$. For any such integral basis we have
$|N(\mu)| = mn$, $mZ = Z \cap \mu\hat{Z}$, and $nZ = \{d : c + d\omega \in \mu\hat{Z}\}$. The
only restriction on s is that $s + n\omega$ be in $\mu\hat{Z}$. Hence m and n, but not
s, are uniquely determined by μ.*
PROOF: Let m and n be the positive rational integers given in
Lemma 4.8, and let $s + n\omega$ be in $\mu\hat{Z}$ ($s + n\omega$ exists by Lemma 4.8).
In view of Lemma 4.9, it suffices to show that $\mu\hat{Z} = \langle m, s + n\omega \rangle$.
Let $c + d\omega$ be in $\mu\hat{Z}$. Since d is in nZ, by Lemma 4.8, then $d = ny$
for some y. Thus since $c + d\omega$ and $s + n\omega$ are in the ideal $\mu\hat{Z}$, we
deduce that $c - sy = c + ny\omega - sy - ny\omega = c + d\omega - y(s + n\omega)$ is
in $Z \cap \mu\hat{Z}$. Consequently, by Lemma 4.8, $c - sy$ in in mZ; and so

$c - sy = mx$ for some x. Hence $c + d\omega = c - sy + y(s + n\omega)$ $= mx + (s + n\omega)y$, and this representation is unique, by Theorem 4.1(iii), because $\Delta(m, s + n\omega) = 5(mn)^2 \neq 0$. Accordingly, $\mu\hat{Z} = \langle m, s + n\omega \rangle$. Q.E.D.

DEFINITION If $\mu \neq 0$ and $\mu\hat{Z} = \langle m, s + n\omega \rangle$, where $m > 0$ and $n > 0$, then $\{m, s + n\omega\}$ is called a **canonical basis** for $\mu\hat{Z}$.

Theorem 4.10 showed that every nonzero ideal $\mu\hat{Z}$ has a canonical basis $\{m, s + n\omega\}$ and that m and n are unique. We now direct our efforts to the actual determination of m, n, and s. The key step in the process is the computation of m and n using the formulas found in the following theorem. Once m and n are determined it is a routine exercise to compute a suitable s. (See example 4.1.)

THEOREM 4.11 *If* $\{m, s + n\omega\}$ *is a canonical basis for the nonzero ideal* $\mu\hat{Z}$, $\mu = a + b\omega$, *and* $d = \gcd(a, b)$, *then we have* $m = |N(\mu)|/d$ *and* $n = d$.
PROOF: Let a_1 and b_1 be given by $a = a_1 d$ and $b = b_1 d$. We have that z is in $Z \cap \mu\hat{Z}$ if and only if there exists $x + y\omega$ such that $z = d(a_1 + b_1\omega)(x + y\omega) = d(a_1 x + b_1 y) + d(b_1 x + (a_1 + b_1)y)\omega$. Consequently, z is in $Z \cap \mu\hat{Z}$ if and only if there exists $x + y\omega$ such that $z = d(a_1 x + b_1 y)$ and $b_1 x + (a_1 + b_1)y = 0$. The general solution of the linear Diophantine equation $b_1 x + (a_1 + b_1)y = 0$ is easily seen to be $x = t(a_1 + b_1)$, $y = -tb_1$, where t is a rational integer parameter. Substituting these values for x and y into $z = d(a_1 x + b_1 y)$, we conclude that z is in $Z \cap \mu\hat{Z}$ if and only if $z = dt(a_1^2 + a_1 b_1 - b_1^2) = dtN(a_1 + b_1\omega) = t(N(a + b\omega)/d)$ $= t(N(\mu)/d)$. Consequently, $Z \cap \mu\hat{Z} = (|N(\mu)|/d)Z$, and since we also have $mZ = Z \cap \mu\hat{Z}$, by Theorem 4.10, we conclude that $m = |N(\mu)|/d$. It then follows that $n = |N(\mu)|/m = d$ from Theorem 4.10. Q.E.D.

EXAMPLE 4.1 We find a canonical basis $\{m, s + n\omega\}$ for the ideal $(4 + 10\omega)\hat{Z}$. Since $\gcd(4, 10) = 2$ and $N(4 + 10\omega) = 16 + 40 - 100 = -44$, it follows from Theorem 4.11 that $m = 44/2 = 22$ and $n = 2$. Thus $(4 + 10\omega)\hat{Z} = \langle 22, s + 2\omega \rangle$, where s is yet to be determined. According to Theorem 4.10, any s such that $s + 2\omega$ is in $(4 + 10\omega)\hat{Z}$ will suffice. From $s + 2\omega = (4 + 10\omega)(x + y\omega)$

$= 4x + 10y + (10x + 14y)\omega$, we see that $s = 4x + 10y$, where $5x + 7y = 1$. Using the usual Euclidean algorithm procedure, we obtain $5(3) + 7(-2) = 1$, so that we may pick $s = 4(3) + 10(-2) = -8$. We therefore have $(4 + 10\omega)\hat{Z} = \langle 22, -8 + 2\omega \rangle$.

V   Residue Class Arithmetic in $\hat{Z}$

Introduction

In this chapter we develop the basic theory of congruence arithmetic in $\hat{Z}$. The development parallels very closely that for rational integer theory, and the following ideas from the latter theory will be generalized to $\hat{Z}$: congruences; Z_n, the ring of integers modulo n; U_n, the group of units of Z_n; the Euler phi-function; complete and reduced residue sets. We will find analogues in $\hat{Z}$ for many of the standard theorems in rational integer theory, including the Euler—Fermat theorem, Fermat's theorem, and the Chinese remainder theorem.

Congruence Notation and Operations on Congruences

For the remainder of the book we adopt the following convention.

CONVENTION   The Greek letter μ denotes a nonzero integer.

We begin the development with the following definitions.

DEFINITIONS   If $\mu \mid (\alpha - \beta)$, we say that the integers α and β are **congruent modulo** μ, and write $\alpha \equiv \beta$ (MOD μ). If $\mu \nmid (\alpha - \beta)$, we say that the integers α and β are **incongruent modulo** μ, and write $\alpha \not\equiv \beta$ (MOD μ). The nonzero integer μ is called the **modulus** of the congruence $\alpha \equiv \beta$ (MOD μ). The set $\{\alpha + \gamma\mu : \gamma \in \hat{Z}\}$ is called the **residue class** of α modulo μ, and is denoted by α^*, or by $\alpha + \mu\hat{Z}$ when the modulus μ is not clearly understood from context. The set of all residue classes modulo μ is denoted by $\hat{Z}_{\mu}$; so that $\hat{Z}_{\mu} = \{\alpha^* : \alpha \in \hat{Z}\}$. Any element of α^* is said to be a **representative** of that residue class.

We have used the notation MOD in the above definitions to avoid any confusion between rational integer residue class arithmetic (mod) and residue class arithmetic in $\hat{Z}$ (MOD). This distinction is somewhat superfluous since if α, β, and μ are rational integers, then $\alpha \equiv \beta$ (mod μ) if and only if $\alpha \equiv \beta$ (MOD μ). However, if one of α, β, or μ is not a rational integer, then rational integer congruence is meaningless even though we may have $\alpha \equiv \beta$ (MOD μ). For example, $5 + 3\omega \equiv 3\omega$ (MOD $2 + \omega$).

Several other observations are appropriate at this time. First, if μ_1 and μ_2 are associates, then it is evident that $\alpha \equiv \beta$ (MOD μ_1) if and only if $\alpha \equiv \beta$ (MOD μ_2). Also, since $\alpha + \mu_1\hat{Z} = \alpha + \mu_2\hat{Z}$, then $\hat{Z}_{\mu_1} = \hat{Z}_{\mu_2}$. However, if α and β are associates, then it does not follow that $\alpha \equiv \beta$ (MOD μ). For example, -1 and 1 are associates but $-1 \not\equiv 1$ (MOD 3). We record these observations in

THEOREM 5.1
(i) *If μ_1 and μ_2 are associates, then $\alpha \equiv \beta$ (MOD μ_1) if and only if $\alpha \equiv \beta$ (MOD μ_2).*
(ii) *If μ_1 and μ_2 are associates, then $\hat{Z}_{\mu_1} = \hat{Z}_{\mu_2}$.*
(iii) *If α and β are associates, then it is not generally true that $\alpha \equiv \beta$ (MOD μ).*

It is evident from the definition of congruence that $\alpha \equiv \beta$ (MOD μ) if and only if $\alpha - \beta$ is in the ideal $\mu\hat{Z}$; and that $\hat{Z}_{\mu} = \hat{Z}/\mu\hat{Z}$. Accordingly, from first principles in ideal theory, we have the following results given in 5.2–5.6.

THEOREM 5.2 *The relation* $\equiv$ *(MOD μ) is an equivalence relation on $\hat{Z}$, and $\hat{Z}_\mu$ is the set of equivalence classes for the relation. Thus the set of residue classes modulo μ is identical to the set of equivalence classes for the relation* $\equiv$ *(MOD μ).*

COROLLARY 5.3
(i) $\alpha \equiv \alpha$ (MOD μ).
(ii) *If* $\alpha \equiv \beta$ (MOD μ), *then* $\beta \equiv \alpha$ (MOD μ).
(iii) *If* $\alpha \equiv \beta$ (MOD μ) *and* $\beta \equiv \gamma$ (MOD μ), *then* $\alpha \equiv \gamma$ (MOD μ).

COROLLARY 5.4
(i) $\alpha^* = \beta^*$ *if and only if* $\alpha \equiv \beta$ (MOD μ).
(ii) *Either* $\alpha^* = \beta^*$ *or else* α^* *and* β^* *are disjoint.*

THEOREM 5.5 $\hat{Z}_\mu$ *is a commutative ring with additive identity* $0^* = \mu\hat{Z}$ *and multiplicative identity* $1^* = 1 + \mu\hat{Z}$ *under the operations* $\alpha^* + \beta^* = (\alpha + \beta)^*$ *and* $\alpha^* \cdot \beta^* = (\alpha\beta)^*$.

DEFINITION $\hat{Z}_\mu$ *is called the* **ring of integers modulo μ**.

COROLLARY 5.6 *If* $\alpha \equiv \beta$ (MOD μ) *and* $\gamma \equiv \delta$ (MOD μ), *then* $\alpha \pm \gamma \equiv \beta \pm \delta$ (MOD μ) *and* $\alpha\gamma \equiv \beta\delta$ (MOD μ).

THEOREM 5.7 *If $f(X)$ is a polynomial with integer coefficients and* $\alpha \equiv \beta$ (MOD μ), *then* $f(\alpha) \equiv f(\beta)$ (MOD μ).
PROOF: If $f(X) = \gamma_0 + \gamma_1 X + \gamma_2 X^2 + \ldots + \gamma_n X^n$, then $f(\alpha) - f(\beta) = \gamma_1(\alpha - \beta) + \gamma_2(\alpha^2 - \beta^2) + \ldots + \gamma_n(\alpha^n - \beta^n)$. For $i = 1, 2, \ldots, n$, we have $\alpha^i - \beta^i = (\alpha - \beta)(\alpha^{i-1} + \alpha^{i-2}\beta + \ldots + \beta^{i-1})$. Thus, since $\mu \mid (\alpha - \beta)$, then $\mu \mid (\alpha^i - \beta^i)$ for $i = 1, 2, \ldots, n$. Consequently, $\mu \mid (f(\alpha) - f(\beta))$, so that $f(\alpha) \equiv f(\beta)$ (MOD μ). Q.E. D.

The next theorem and its corollary will be used many times.

THEOREM 5.8 *If $\delta = \text{GCD}(\gamma, \mu)$, then* $\alpha\gamma \equiv \beta\gamma$ (MOD μ) *if and only if* $\alpha \equiv \beta$ (MOD μ/δ).
PROOF: There exist γ_1 and μ_1 such that $\gamma = \delta\gamma_1$, $\mu = \delta\mu_1$,

and $GCD(\gamma_1, \mu_1) = 1$. Since $\mu_1 = \mu/\delta$, we are to show that $\alpha\gamma \equiv \beta\gamma$ (MOD μ) if and only if $\alpha \equiv \beta$ (MOD μ_1). If $\alpha\gamma \equiv \beta\gamma$ (MOD μ), then $\delta\mu_1 | (\alpha - \beta)\delta\gamma_1$; so that $\mu_1 | (\alpha - \beta)\gamma_1$. Since $GCD(\gamma_1, \mu_1) = 1$, it follows from Euclid's lemma that $\mu_1 | (\alpha - \beta)$. Thus $\alpha \equiv \beta$ (MOD μ_1). The converse follows from the fact that if $\mu_1 | (\alpha - \beta)$, then $\delta\mu_1 | (\alpha - \beta)\delta\gamma_1$. Q.E.D.

COROLLARY 5.9 *If* $GCD(\gamma, \mu) = 1$, *then* $\alpha\gamma \equiv \beta\gamma$ (MOD μ) *if and only if* $\alpha \equiv \beta$ (MOD μ).

Complete Residue Sets

The following theorem is of fundamental importance. From it we deduce that $\hat{Z}_\mu$ is a finite commutative ring with $|N(\mu)|$ elements. Moreover, an explicit set of representatives for the $|N(\mu)|$ residue classes is given.

THEOREM 5.10 *If* $\{m, s + n\omega\}$ *is a canonical basis for* $\mu\hat{Z}$, *then* $\hat{Z}_\mu$ *has exactly* $|N(\mu)| = mn$ *elements and they are given by* $(r_1 + r_2\omega)^*$, *where* $0 \leq r_1 < m$ *and* $0 \leq r_2 < n$.
PROOF: By Theorem 4.10, we have $|N(\mu)| = mn$, $mZ = Z \cap \mu\hat{Z}$, and $nZ = \{d : c + d\omega \in \mu\hat{Z}\}$. If $(r_1 + r_2\omega)^* = (r_1' + r_2'\omega)^*$, then $r_1 - r_1' + (r_2 - r_2')\omega$ is in μZ. Thus, because $n | (r_2 - r_2')$ and $|r_2 - r_2'| < n$, then $r_2 = r_2'$. Consequently, $r_1 - r_1'$ is in $\mu\hat{Z}$. Thus $m | (r_1 - r_1')$ and $|r_1 - r_1'| < m$, so that $r_1 = r_1'$. Hence the mn elements $(r_1 + r_2\omega)^*$ of $\hat{Z}_\mu$ are distinct. All that remains to be shown is that if $c + d\omega$ is in $\hat{Z}$, then $(c + d\omega)^* = (r_1 + r_2\omega)^*$ for suitable r_1 and r_2 satisfying $0 \leq r_1 < m$ and $0 \leq r_2 < n$. By the division algorithm for rational integers, we have $d = q_1 n + r_2$, $0 \leq r_2 < n$, and $c - q_1 s = q_2 m + r_1$, $0 \leq r_1 < m$. Thus

$$(c + d\omega)^* = (q_1 s + q_2 m + r_1 + (q_1 n + r_2)\omega)^*$$
$$= (r_1 + r_2\omega + (q_2 m + q_1(s + n\omega)))^*$$
$$= (r_1 + r_2\omega)^*,$$

where the last step follows from the fact that $q_2 m + q_1(s + n\omega)$ is in $\mu \hat{Z}$. Q.E.D.

DEFINITION Let $t = |N(\mu)|$. If $\{\xi_1, \xi_2, \ldots, \xi_t\}$ is a set consisting of exactly one element from each of the t residue classes modulo μ, then $\{\xi_1, \xi_2, \ldots, \xi_t\}$ is called a **complete set of residues modulo μ**.

THEOREM 5.11 *If* $\{\xi_1, \xi_2, \ldots, \xi_t\}$ *is a complete set of residues modulo μ, then*
(i) $\hat{Z}_\mu = \{\xi_1{}^*, \xi_2{}^*, \ldots, \xi_t{}^*\}$;
(ii) *if μ and μ_1 are associates, then* $\{\xi_1, \xi_2, \ldots, \xi_t\}$ *is a complete set of residues modulo μ_1.*
PROOF: Assertion (i) is immediate from the definition of complete set of residues and assertion (ii) is almost as easy. For, by Theorem 5.1(ii), $\hat{Z}_\mu = \hat{Z}_{\mu_1}$. Thus $\{\xi_1{}^*, \xi_2{}^*, \ldots, \xi_t{}^*\}$ is the set of t distinct residue classes modulo μ_1 and, consequently, $\{\xi_1, \xi_2, \ldots, \xi_t\}$ is a set consisting of exactly one element from each of the t residue classes modulo μ_1. Q.E.D.

THEOREM 5.12 *If* $t = |N(\mu)|$, *then* $\{\xi_1, \xi_2, \ldots, \xi_t\}$ *is a complete set of residues modulo μ if and only if $\xi_i \not\equiv \xi_j$ (MOD μ) whenever $i \neq j$.*
PROOF: Each element of the set $\{\xi_1, \xi_2, \ldots, \xi_t\}$ comes from a different residue class modulo μ if and only if $\xi_i \not\equiv \xi_j$ (MOD μ) whenever $i \neq j$. Since there are $t = |N(\mu)|$ residue classes modulo μ, the desired result follows. Q.E.D.

The following example illustrates how Theorems 4.11 and 5.10 may be used to determine a complete set of residues.

EXAMPLE 5.1 Let $\{m, s + n\omega\}$ be a canonical basis for $\mu\hat{Z}$, where $\mu = 2 + 6\omega$. Since $\gcd(2, 6) = 2$ and $N(2 + 6\omega) = 4 + 12 - 36 = -20$, it follows from Theorem 4.11 that $n = 2$ and $m = 20/2 = 10$. Thus it follows from Theorem 5.10 that $\{0, 1, 2, 3, 4, 5, 6, 7, 8, 9, \omega, 1 + \omega, 2 + \omega, 3 + \omega, 4 + \omega, 5 + \omega, 6 + \omega, 7 + \omega, 8 + \omega, 9 + \omega\}$ is a complete set of residues modulo $2 + 6\omega$.

If ξ is given, then we may apply the division algorithm for $\hat{Z}$ to obtain $\xi = \mu\theta + \rho$, where $|N(\rho)| < |N(\mu)|$. Consequently, $\xi \equiv \rho$ (MOD μ) and $|N(\rho)| < |N(\mu)|$. Hence the division algorithm for $\hat{Z}$ allows us to find a complete set of residues modulo μ in which the norm of each element of the set has absolute value less than $|N(\mu)|$. (Note the analogy with rational integer number theory.)

EXAMPLE 5.2 In Example 5.1 the norms of the elements $0, 1, 2,$ $3, 4, \omega, 1 + \omega, 2 + \omega, 3 + \omega,$ and $4 + \omega$ have absolute values less than $|N(2 + 6\omega)| = 20$. By the division algorithm, we obtain $5 = (2 + 6\omega) \cdot (-2 + \omega) + (3 + 4\omega)$, $|N(3 + 4\omega)| = 5$. Thus $5 \equiv 3 + 4\omega$ (MOD $2 + 6\omega$) and $|N(3 + 4\omega)| < 20$. Similarly, modulo $2 + 6\omega$, we obtain $6 \equiv -2 - 4\omega$, $7 \equiv 1 + 2\omega$, $8 \equiv 2 + 2\omega$, $9 \equiv -1$, $5 + \omega$ $\equiv 3 + 5\omega$, $6 + \omega \equiv -2 - 3\omega$, $7 + \omega \equiv 1 + 3\omega$, $8 + \omega \equiv 2 + 3\omega$, and $9 + \omega \equiv -3 - 5\omega$, where the norm of the right side of each congruence has absolute value less than 20. Consequently, $\{0, 1, 2, 3, 4, \omega,$ $1 + \omega, 2 + \omega, 3 + \omega, 4 + \omega, 3 + 4\omega, -2 - 4\omega, 1 + 2\omega, 2 + 2\omega, -1,$ $3 + 5\omega, -2 - 3\omega, 1 + 3\omega, 2 + 3\omega, -3 - 5\omega\}$ is a complete set of residues modulo $2 + 6\omega$ in which the norm of each element of the set has absolute value less than $20 = |N(2 + 6\omega)|$.

Complete Residue Sets for Particular Moduli

In the next three theorems we determine complete residue sets for various moduli μ. The procedure in each case is similar to the technique used in Example 5.1, where we used Theorems 4.11 and 5.10. We shall also use the result given in Theorem 5.11: If μ and μ_1 are associates, then a complete set of residues modulo μ is also a complete set of residues modulo μ_1. The numbers m and n used in the proofs of the three theorems are those given in Theorem 4.11.

THEOREM 5.13 *If* $a \neq 0$, *then* $\{r_1 + r_2\omega : 0 \leq r_1 < |a|,\ 0 \leq r_2 < |a|\}$ *is a complete set of residues modulo a.*

PROOF: Since $a = a + 0\omega$, $\gcd(a, 0) = |a|$, and $N(a) = a^2$, then $n = |a|$ and $m = a^2/|a| = |a|$. Consequently, $\{r_1 + r_2\omega : 0 \leq r_1 < |a|, 0 \leq r_2 < |a|\}$ is a complete set of residues modulo a. Q.E.D.

THEOREM 5.14 *If* $k > 0$, q *is a rational prime of the form* $q \equiv \pm 1 \pmod{5}$, *and* π *is a prime such that* $|N(\pi)| = q$, *then* $\{0, 1, 2, \ldots, q^k - 1\}$ *is a complete set of residues modulo* π^k.

PROOF: Suppose $a + b\omega = \pi^k$. If $\gcd(a, b) \neq 1$, then there exists a rational prime which divides a and b; and so, this rational prime also divides $\pi^k = a + b\omega$. This rational prime must be q, because of Theorem 3.5. But since $\bar{\pi} \mid q$, we would then have $\bar{\pi} \mid \pi^k$, and so $\bar{\pi} \mid \pi$. But this is impossible, as π and $\bar{\pi}$ are not associates, by Lemma 3.9. Thus $\gcd(a, b) = 1$; and so $n = 1$ and $m = |N(a + b\omega)| = |N(\pi^k)| = q^k$. Therefore $\{0, 1, 2, \ldots, q^k - 1\}$ is a complete set of residues modulo π^k. Q.E.D.

THEOREM 5.15 *If* $k > 0$, *then*

(i) $\{r_1 + r_2\omega : 0 \leq r_1 < 5^k, 0 \leq r_2 < 5^k\}$ *is a complete set of residues modulo* $(2 + \omega)^{2k}$;

(ii) $\{r_1 + r_2\omega : 0 \leq r_1 < 5^k, 0 \leq r_2 < 5^{k-1}\}$ *is a complete set of residues modulo* $(2 + \omega)^{2k-1}$.

PROOF: Since $(2 + \omega)^2 = 5\omega^2$, then $(2 + \omega)^{2k} = 5^k\omega^{2k}$. Therefore $(2 + \omega)^{2k}$ is an associate of 5^k, and it follows from Theorem 5.13 that the set in (i) is a complete set of residues modulo $(2 + \omega)^{2k}$. In order to establish (ii), we observe that $(2 + \omega)^{2k-1} = (2 + \omega)^{2k-2}(2 + \omega) = \omega^{2k-2}(5^{k-1})(2 + \omega)$. Consequently, $(2 + \omega)^{2k-1}$ is an associate of $5^{k-1}(2 + \omega)$. Applying our now familiar argument, we obtain $n = 5^{k-1}$ and $m = |N(5^{k-1}(2 + \omega))|/5^{k-1} = 5^k$. Thus the set given in (ii) is a complete set of residues modulo $(2 + \omega)^{2k-1}$, and we are done. Q.E.D.

Further Results on Complete Residue Sets

Given the complete set of residues $\{r_1 + r_2\omega : 0 \leq r_1 < m, 0 \leq r_2 < n\}$ modulo μ furnished by Theorem 5.10, and an arbitrary α, we can use the procedure given in the proof to determine the $r_1 + r_2\omega$ in the set such that $\alpha \equiv r_1 + r_2\omega$ (MOD μ). In the case that μ is a prime, there are easier ways to make this determination. Since in many applications μ will be a prime, it will be well worth the effort to simplify the work involved. The analysis separates into two cases: (i) π is an associate of a rational prime p of the form $p \equiv \pm 2$ (mod 5); (ii) $|N(\pi)| = q$, where q is a rational prime. The former case is trivial to handle, for then $\{r_1 + r_2\omega : 0 \le r_1 < p, 0 \le r_2 < p\}$ is a complete set of residues modulo π, and it is plain that $a + b\omega \equiv r_1 + r_2\omega$ (MOD π) if and only if $a \equiv r_1$ (mod p) and $b \equiv r_2$ (mod p). In the latter case, $\{0, 1, 2, \ldots, q - 1\}$ is a complete set of residues modulo π, and, given an arbitrary $a + b\omega$, a little more effort will be required to find the k, $0 \le k < q$, such that $a + b\omega \equiv k$ (MOD π). The following theorem is useful toward this end.

THEOREM 5.16 *If q is a rational prime and π is a prime such that $|N(\pi)| = q$, then $a \equiv b$ (MOD π) if and only if $a \equiv b$ (mod q).*
PROOF: If $a \equiv b$ (mod q), then $a \equiv b$ (MOD $\pi\bar{\pi}$); and so $a \equiv b$ (MOD π). Conversely, if $a \equiv b$ (MOD π), then $\pi | (a - b)$. If $a - b = 0$, then we certainly have $a \equiv b$ (mod q). Thus we may suppose that $a - b = \pm p_1 p_2 \cdots p_r$, where the p_i are rational primes. Since $\pi | (a - b)$, then π must divide one of the p_i, by Theorem 2.21. But, by Theorem 3.5, q is the only rational prime divisible by π. Consequently, q is a factor of $a - b$, and so $a \equiv b$ (mod q). Q.E.D.

Continuing now our analysis of Case (ii), let $\pi = c + d\omega$. Then $|N(c + d\omega)| = q$, so that $c^2 + cd - d^2 = \pm q$. From this equation we deduce that c and d are relatively prime to q, for otherwise we would be led to the impossible situation that $q^2 | q$. Thus $\gcd(d, q) = 1$, and so we can determine a t such that $dt \equiv 1$ (mod q). Using Theorem 5.16 and $c + d\omega \equiv 0$ (MOD π), we obtain that $\omega \equiv (dt)\omega \equiv -ct$ (MOD π). Consequently, if $a + b\omega$ is any integer, then $a + b\omega \equiv a + b(-ct)$ (MOD π). The right side of the last congruence is a rational integer and, by another application of Theorem 5.16, is congruent modulo π to a rational integer in the set $\{0, 1, 2, \ldots, q - 1\}$. We illustrate these ideas via an example.

EXAMPLE 5.3 $\{0, 1, 2, \ldots, 540\}$ is a complete set of residues modulo $22 + 3\omega$, since 541 is a rational prime and $|N(22 + 3\omega)|$ $= 541$. Since $541 = 3(180) + 1$, then $3(-180) \equiv 1 \pmod{541}$. Consequently, $\omega \equiv 3(-180)\omega \equiv (-180)(-22) = 3960 \equiv 173$ (MOD $22 + 3\omega$). Then, for example, we have $5 + 3\omega \equiv 5 + 3(173)$ $= 524$ (MOD $22 + 3\omega$), and $39 - 12\omega \equiv 39 - 12(173) = -2037$ $\equiv 127$ (MOD $22 + 3\omega$).

THEOREM 5.17
(i) $\alpha \equiv \beta$ (MOD μ) *if and only if* $\bar{\alpha} \equiv \bar{\beta}$ (MOD $\bar{\mu}$).
(ii) $\{\xi_1, \xi_2, \ldots, \xi_t\}$ *is a complete set of residues modulo* μ *if and only if* $\{\bar{\xi_1}, \bar{\xi_2}, \ldots, \bar{\xi_t}\}$ *is a complete set of residues modulo* $\bar{\mu}$.
PROOF: If $\mu | (\alpha - \beta)$, then $\alpha - \beta = \mu\delta$; and so $\bar{\alpha} - \bar{\beta} = \bar{\mu}\bar{\delta}$. Thus $\bar{\mu} | (\bar{\alpha} - \bar{\beta})$. Conversely, if $\bar{\mu} | (\bar{\alpha} - \bar{\beta})$, then a similar argument shows that $\mu | (\alpha - \beta)$. Accordingly, $\mu | (\alpha - \beta)$ if and only if $\bar{\mu} | (\bar{\alpha} - \bar{\beta})$. This establishes (i).

From (i) we conclude that the ξ_i are incongruent modulo μ if and only if the $\bar{\xi_i}$ are incongruent modulo $\bar{\mu}$. Since $t = |N(\mu)|$ $= |N(\bar{\mu})|$, the result now follows from Theorem 5.12. Q.E.D.

THEOREM 5.18 *If* $\{\xi_1, \xi_2, \ldots, \xi_t\}$ *is a complete set of residues modulo* μ, $\text{GCD}(\alpha, \mu) = 1$, *and* γ *is any integer, then* $\{\alpha\xi_1 + \gamma, \alpha\xi_2 + \gamma, \ldots, \alpha\xi_t + \gamma\}$ *is a complete set of residues modulo* μ.
PROOF: Suppose $\alpha\xi_i + \gamma \equiv \alpha\xi_j + \gamma$ (MOD μ). Then $\alpha\xi_i \equiv \alpha\xi_j$ (MOD μ); and since $\text{GCD}(\alpha, \mu) = 1$, we obtain $\xi_i \equiv \xi_j$ (MOD μ). Consequently, $i = j$. Hence the $t = |N(\mu)|$ elements $\alpha\xi_i + \gamma$ are incongruent modulo μ, and therefore constitute a complete set of residues modulo μ. Q.E.D.

THEOREM 5.19 *If* $\text{GCD}(\mu_1, \mu_2) = 1$, $\{\xi_1, \xi_2, \ldots, \xi_t\}$ *is a complete set of residues modulo* μ_1, *and* $\{\eta_1, \eta_2, \ldots, \eta_s\}$ *is a complete set of residues modulo* μ_2, *then* $\{\mu_2\xi_i + \mu_1\eta_j : i = 1, 2, \ldots, t, \ j = 1, 2, \ldots, s\}$ *is a complete set of residues modulo* $\mu_1\mu_2$.
PROOF: Suppose $\mu_2\xi_i + \mu_1\eta_j \equiv \mu_2\xi_h + \mu_1\eta_k$ (MOD $\mu_1\mu_2$). Then $\mu_2(\xi_i - \xi_h) + \mu_1(\eta_j - \eta_k) \equiv 0$ (MOD $\mu_1\mu_2$), and thus $\mu_2(\xi_i - \xi_h) \equiv 0$ (MOD μ_1) and $\mu_1(\eta_j - \eta_k) \equiv 0$ (MOD μ_2). As $\text{GCD}(\mu_1, \mu_2) = 1$, we conclude that $\xi_i \equiv \xi_h$ (MOD μ_1) and $\eta_j \equiv \eta_k$ (MOD μ_2). Accordingly, $i = h$ and $j = k$. Thus the $|N(\mu_1\mu_2)| = |N(\mu_1)| \cdot |N(\mu_2)| = st$ elements $\mu_2\xi_i + \mu_1\eta_j$ are incongruent modulo $\mu_1\mu_2$, and therefore constitute a complete set of residues modulo $\mu_1\mu_2$. Q.E.D.

EXAMPLE 5.4 $|N(1 + 3\omega)| = 5$, and so $1 + 3\omega$ is an associate of $2 + \omega$. It follows, by Theorem 5.15(ii), that $\{0, 1, 2, 3, 4\}$ is a complete set of residues modulo $1 + 3\omega$. Also, by Theorem 5.13, $\{0, 1, \omega, 1 + \omega\}$ is a complete set of residues modulo 2. Since $GCD(2, 1 + 3\omega) = 1$, we conclude from Theorem 5.19 that $\{(1 + 3\omega)(r_1 + r_2\omega) + 2k : 0 \le r_1 \le 1, \; 0 \le r_2 \le 1, \; 0 \le k \le 4\}$ is a complete set of residues modulo $2 + 6\omega$. Listing the members of this set in the same order modulo $2 + 6\omega$ as the elements of the set found in Example 5.1, we obtain the complete set of residues $\{0, 9 + 4\omega, 2, 11 + 4\omega, 4, 3 + 4\omega, 6, 5 + 4\omega, 8, 7 + 4\omega, 12 + 7\omega, 5 + 3\omega, 4 + 7\omega, 7 + 3\omega, 6 + 7\omega, 9 + 3\omega, 8 + 7\omega, 1 + 3\omega, 10 + 7\omega, 3 + 3\omega\}$. For example, the sixth element, 5, of the set in Example 5.1 is congruent modulo $2 + 6\omega$ to the sixth element, $3 + 4\omega$, of the above set.

THEOREM 5.20 *If α and β are in the same residue class modulo μ, then $GCD(\alpha, \mu) = GCD(\beta, \mu)$.*
PROOF: Since $\alpha \equiv \beta$ (MOD μ), then $\alpha = \mu\gamma + \beta$. It then follows from Theorem 2.18(iv) that $GCD(\alpha, \mu) = GCD(\beta + \mu\gamma, \mu) = GCD(\beta, \mu)$. Q.E.D.

COROLLARY 5.21 *If $GCD(\alpha, \mu) = 1$ and β is in α^*, then $GCD(\beta, \mu) = 1$. Thus, if any element in a residue class modulo μ is relatively prime to μ, then all elements in that residue class are.*

$\hat{U}_\mu$, *the Group of Units of* $\hat{Z}_\mu$

NOTATION $\hat{U}_\mu = \{\alpha^* \in \hat{Z}_\mu : GCD(\alpha, \mu) = 1\}$.

It follows from Corollary 5.21 that $\hat{U}_\mu$ consists of those residue classes modulo μ whose elements are relatively prime to μ.

THEOREM 5.22 *An element α^* in $\hat{Z}_\mu$ has a multiplicative inverse* *if and only if α^* is in $\hat{U}_\mu$. Moreover, 0^* is in $\hat{U}_\mu$ if and only if μ is a unit of $\hat{Z}$. Consequently, 0^* has a multiplicative inverse in $\hat{Z}_\mu$ if and only if $\hat{Z}_\mu$ is the trivial ring $\hat{Z}_\mu = \{0^*\}$.*

PROOF: If α^* is in $\hat{U}_\mu$, then $\text{GCD}(\alpha, \mu) = 1$. Accordingly, there exist β and γ such that $\alpha\beta + \mu\gamma = 1$. Hence, $1^* = (\alpha\beta + \mu\gamma)^* = (\alpha\beta)^* = \alpha^* \cdot \beta^*$, and so β^* is the multiplicative inverse of α^*.

Conversely, if α^* has a multiplicative inverse, then there exists a β such that $\alpha^* \cdot \beta^* = 1^*$. Then $(\alpha\beta)^* = 1^*$, and so $\mu \,|\, (\alpha\beta - 1)$. But then $\alpha\beta - 1 = \mu\gamma$, so that $\alpha\beta - \mu\gamma = 1$. Thus we see that any common divisor of α and μ must divide 1; that is, any common divisor of α and μ is a unit. Accordingly, $\text{GCD}(\alpha, \mu) = 1$, so that α^* is in $\hat{U}_\mu$.

If μ is a unit, then $\text{GCD}(0, \mu) = 1$. Thus 0^* is in $\hat{U}_\mu$, and so has a multiplicative inverse. Conversely, if 0^* is in $\hat{U}_\mu$, then $\text{GCD}(0, \mu) = 1$, and we deduce that μ is a unit. The last assertion of the theorem follows from that fact that $\hat{Z}_\mu = \{0^*\}$ if and only if μ is a unit. Q.E.D.

An immediate consequence of Theorem 5.22 is the important

COROLLARY 5.23 *$\hat{U}_\mu$ is the finite multiplicative abelian group of units of $\hat{Z}_\mu$. The elements of $\hat{U}_\mu$ are the residue classes of $\hat{Z}_\mu$ consisting of integers that are relatively prime to μ.*

LEMMA 5.24 *If $1 \leq |N(\alpha)| < |N(\mu)|$, then α is not in $\mu\hat{Z}$, and so $\alpha^* \neq 0^*$.*

PROOF: Note that $\alpha \neq 0$ since $|N(\alpha)| \geq 1$. Suppose, contrariwise, that α is in $\mu\hat{Z}$. Then $\alpha = \mu\gamma$ for some $\gamma \neq 0$, so that $|N(\alpha)| = |N(\mu\gamma)| = |N(\mu)| \cdot |N(\gamma)| \geq |N(\mu)|$, a contradiction. Q.E.D.

THEOREM 5.25 *The zero divisors of $\hat{Z}_\mu$ are precisely the nonzero elements α^* which are not elements of $\hat{U}_\mu$.*

PROOF: If $\alpha^* \neq 0^*$ is not in $\hat{U}_\mu$, then $\text{GCD}(\alpha, \mu) = \delta$, where $\delta \neq 0$ is not a unit in $\hat{Z}$. Therefore $\alpha = \alpha_1\delta$ and $\mu = \mu_1\delta$ for some α_1 and μ_1, and since $|N(\mu)| > |N(\mu_1)| \geq 1$, it follows from Lemma 5.24 that $\mu_1{}^* \neq 0^*$. Consequently, we have $\alpha^* \cdot \mu_1{}^* = (\alpha\mu_1)^* = (\alpha_1\mu_1\delta)^* = (\alpha_1\mu)^* = \alpha_1{}^* \cdot \mu^* = \alpha_1{}^* \cdot 0^* = 0^*$, so that α^* is a zero divisor of $\hat{Z}_\mu$.

Conversely, suppose $\alpha^* \neq 0^*$ is a zero divisor of $\hat{Z}_\mu$. Then there exists a $\beta^* \neq 0^*$ such that $\alpha^* \cdot \beta^* = 0^*$. Now if α^* were to belong to $\hat{U}_\mu$, then α^* would have a multiplicative inverse γ^*. Then $\beta^* = 1^* \cdot \beta = (\gamma^* \cdot \alpha^*)\beta^* = \gamma^*(\alpha^* \cdot \beta^*) = \gamma^* \cdot 0^* = 0^*$, contradicting the fact that $\beta^* \neq 0^*$. Accordingly, α^* is not in $\hat{U}_\mu$. Q.E.D.

THEOREM 5.26 $\hat{Z}_\mu$ *is a field if and only if μ is a prime.*
PROOF: First suppose that μ is a prime. We must show that $\hat{Z}_\mu$ has at least two elements, and every $\alpha^* \neq 0^*$ has a multiplicative inverse. Since μ is a prime, then $\text{GCD}(\alpha, \mu) = 1$ for every α not in $\mu\hat{Z} = 0^*$. Consequently, α^* is in $\hat{U}_\mu$ for every $\alpha^* \neq 0^*$, and so every nonzero α^* has a multiplicative inverse. In addition, $\hat{Z}_\mu$ has $|\text{N}(\mu)| > 1$ elements. Hence, if μ is a prime, then $\hat{Z}_\mu$ is a field.

Next suppose that $\hat{Z}_\mu$ is a field. Since $\hat{Z}_\mu$ has $|\text{N}(\mu)| \geq 2$ elements, μ cannot be a unit. If, in addition, μ is not a prime, then $\mu = \alpha\beta$ for some nonzero elements α and β, neither of which is a unit. This means $1 < |\text{N}(\alpha)| < |\text{N}(\mu)|$ and $1 < |\text{N}(\beta)| < |\text{N}(\mu)|$; hence that $\alpha^* \neq 0^*$ and $\beta^* \neq 0^*$, by Lemma 5.24. Furthermore, $\alpha^* \cdot \beta^* = (\alpha\beta)^* = \mu^* = 0^*$. Hence $\alpha^* \neq 0^*$ is a zero divisor of $\hat{Z}_\mu$. Thus, by Theorem 5.25, α^* is not in $\hat{U}_\mu$. Hence, by Corollary 5.23, $\alpha^* \neq 0^*$ does not have a multiplicative inverse. This contradicts the fact that $\hat{Z}_\mu$ is a field, and so μ must be prime. Q.E.D.

We could also have deduced Theorem 5.26 directly from Theorem 2.23 and the following result in ring theory: In a commutative ring R with identity, a proper ideal M of R is a maximal ideal if and only if R/M is a field.

The Φ Function and the Theorems of Euler and Fermat

DEFINITION For each nonzero μ, $\Phi(\mu)$ denotes the number of residue classes in $\hat{U}_\mu$. The function Φ is called the **Euler phi-function** (or sometimes, the **totient function**) for $\hat{Z}$.

From the preceding definition and Corollary 5.23, we observe that $\Phi(\mu)$ is the order of $\hat{U}_\mu$, the finite multiplicative abelian group of units of $\hat{Z}_\mu$. Some simple properties of the totient function for $\hat{Z}$ are listed in

THEOREM 5.27
(i) $\Phi(\mu) \leq |N(\mu)|$, *with equality if and only if μ is a unit.*
(ii) *If μ is a unit, then $\Phi(\mu) = 1$.*
(iii) *If μ and μ_1 are associates, then $\Phi(\mu) = \Phi(\mu_1)$.*
(iv) *If μ is a prime, then $\Phi(\mu) = |N(\mu)| - 1$.*
PROOF: Statements (i) and (ii) follow from Theorem 5.22; (iii) follows from the fact that $\hat{Z}_\mu = \hat{Z}_{\mu_1}$; and (iv) follows from Theorem 5.26. Q.E.D.

Analogous to rational integer theory we have

THEOREM 5.28 (Euler–Fermat theorem for $\hat{Z}$) *If* $\mathrm{GCD}(\alpha, \mu) = 1$, *then* $\alpha^{\Phi(\mu)} \equiv 1$ (MOD μ).
PROOF: If G is any group of order n and $g \in G$, then $g^n = 1$. Since $\Phi(\mu)$ is the order of the group $\hat{U}_\mu$ and α^* is in $\hat{U}_\mu$, it follows that $(\alpha^{\Phi(\mu)})^* = (\alpha^*)^{\Phi(\mu)} = 1^*$. Consequently, $\alpha^{\Phi(\mu)} \equiv 1$ (MOD μ). Q.E.D.

Later in this chapter we shall give another proof of Theorem 5.28. A simple consequence of Theorem 5.28 is

THEOREM 5.29 (Fermat's theorem for $\hat{Z}$) *If π is a prime and α is an integer, then*
(i) $\alpha^{|N(\pi)|} \equiv \alpha$ (MOD π);
(ii) *if* $\mathrm{GCD}(\alpha, \pi) = 1$, *then* $\alpha^{|N(\pi)|-1} \equiv 1$ (MOD π).
PROOF: If $\mathrm{GCD}(\alpha, \pi) = 1$, then $\alpha^{|N(\pi)|-1} = \alpha^{\Phi(\pi)} \equiv 1$ (MOD π), by the Euler–Fermat theorem. Multiplying both sides of the last congruence by α, we obtain $\alpha^{|N(\pi)|} \equiv \alpha$ (MOD π). Finally, if $\mathrm{GCD}(\alpha, \pi) \neq 1$, then $\pi \mid \alpha$. In this case $\alpha \equiv 0$ (MOD π), and so $\alpha^{|N(\pi)|} \equiv 0 \equiv \alpha$ (MOD π). Q.E.D.

We have used Φ to denote the totient function for $\hat{Z}$. We will use ϕ to denote the totient function for rational integer theory; that is, if $n > 0$, then $\phi(n)$ is the number of positive rational integers less than or equal to n that are relatively prime to n. It is interesting to compare the values of $\Phi(n)$ and $\phi(n)$ when $n > 0$. Of course, $\Phi(1) = \phi(1) = 1$. However, as we shall see in Corollary 5.45, if $n > 1$, then $\Phi(n) > \phi(n)$. Indeed, it will be shown in Theorem 5.44 that $\Phi(n) \geq (\phi(n))^2$ whenever $n > 1$.

Reduced Residue Sets

DEFINITION The set $\{\xi_1, \xi_2, \ldots, \xi_t\}$, where $t = \Phi(\mu)$, is called a **reduced set of residues modulo μ** if $\{\xi_1, \xi_2, \ldots, \xi_t\}$ consists of exactly one element from each of the $\Phi(\mu)$ residue classes in $\hat{U}_\mu$.

THEOREM 5.30 *Let $\{\xi_1, \xi_2, \ldots, \xi_t\}$ be a reduced set of residues modulo μ. Then*
(i) $\hat{U}_\mu = \{\xi_1{}^*, \xi_2{}^*, \ldots, \xi_t{}^*\}$;
(ii) *if μ and μ_1 are associates, then $\{\xi_1, \xi_2, \ldots, \xi_t\}$ is a reduced set of residues modulo μ_1.*
PROOF: Similar to that of Theorem 5.11. Q.E.D.

THEOREM 5.31 *If $t = \Phi(\mu)$, then $\{\xi_1, \xi_2, \ldots, \xi_t\}$ is a reduced set of residues modulo μ if and only if $GCD(\xi_i, \mu) = 1$ for each i and $\xi_i \not\equiv \xi_j \, (MOD \, \mu)$ whenever $i \neq j$.*
PROOF: Similar to that of Theorem 5.12. Q.E.D.

A Preliminary Discussion of the Problem of Evaluating $\Phi(\mu)$

We now direct our attention to developing a computational formula for $\Phi(\mu)$. The final result is analogous to the formula for

$\phi(m)$: If $\pi_1, \pi_2, \ldots, \pi_n$ are the nonassociated prime factors of μ, then

$$\Phi(\mu) = |N(\mu)| \cdot \prod_{i=1}^{n} \left(1 - |\mathrm{N}(\pi_i)|^{-1}\right). \tag{5.1}$$

The procedure whereby Formula (5.1) is obtained is the same as for the ϕ function. First we establish that if $k > 0$ and π is a prime, then

$$\Phi(\pi^k) = |\mathrm{N}(\pi^k)| \cdot \left(1 - |\mathrm{N}(\pi)|^{-1}\right). \tag{5.2}$$

Next we show that $\Phi(\mu_1 \mu_2) = \Phi(\mu_1) \cdot \Phi(\mu_2)$ when $\mathrm{GCD}(\mu_1, \mu_2) = 1$. Formula (5.1) is then an easy consequence of this result and (5.2).

In order to establish (5.2), we consider the complete residue sets obtained in Theorems 5.13, 5.14, and 5.15 and determine those elements in each set that are not relatively prime to the modulus. The remaining elements in each set then form a reduced set of residues for that modulus, and Formula (5.2) will follow in short order.

We will develop the Chinese remainder theorem for $\hat{Z}$ in order to deduce the result $\Phi(\mu_1 \mu_2) = \Phi(\mu_1) \cdot \Phi(\mu_2)$ when $\mathrm{GCD}(\mu_1, \mu_2) = 1$. Alternate methods avoiding the Chinese remainder theorem are available, but the ease with which the result follows from this theorem makes its early introduction appropriate.

Evaluation of $\Phi(\pi^k)$

We base the derivation of (5.2) on the series of seven lemmas given in 5.32–5.38. As will be evident from the presentation, the most troublesome case is when π is an associate of $2 + \omega$.

LEMMA 5.32 $(2 + \omega)|(r_1 + r_2\omega)$ *if and only if* $r_1 \equiv 2r_2 \pmod 5$.
PROOF: We have

$$(r_1 + r_2\omega)/(2 + \omega) = (1/5)(3r_1 - r_2) + (1/5)(2r_2 - r_1)\omega.$$

Thus $(2 + \omega)|(r_1 + r_2\omega)$ if and only if $3r_1 \equiv r_2 \pmod 5$ and $r_1 \equiv 2r_2 \pmod 5$. Since $3r_1 \equiv r_2 \pmod 5$ if and only if $r_1 \equiv 2r_2 \pmod 5$, the desired result follows. Q.E.D.

LEMMA 5.33 *If $k > 0$, then the integers in the set $\{r_1 + r_2\omega : 0 \le r_1 < 5^k, 0 \le r_2 < 5^{k-1}\}$ of complete residues modulo $(2 + \omega)^{2k-1}$ that are divisible by $2 + \omega$ are*
(i) $5j + r_2\omega$, $0 \le j < 5^{k-1}$, *when* $r_2 \equiv 0 \pmod 5$;
(ii) $2 + 5j + r_2\omega$, $0 \le j < 5^{k-1}$, *when* $r_2 \equiv 1 \pmod 5$;
(iii) $4 + 5j + r_2\omega$, $0 \le j < 5^{k-1}$, *when* $r_2 \equiv 2 \pmod 5$;
(iv) $1 + 5j + r_2\omega$, $0 \le j < 5^{k-1}$, *when* $r_2 \equiv 3 \pmod 5$;
(v) $3 + 5j + r_2\omega$, $0 \le j < 5^{k-1}$, *when* $r_2 \equiv 4 \pmod 5$.
PROOF: An easy consequence of Lemma 5.32. Q.E.D.

LEMMA 5.34 *If $k > 0$ and $|N(\pi)| = 5$, then*
$$\Phi(\pi^{2k-1}) = |N(\pi^{2k-1})| \cdot \left(1 - |N(\pi)|^{-1}\right).$$

PROOF: The prime π is an associate of $2 + \omega$ and thus $\Phi(\pi^{2k-1}) = \Phi((2 + \omega)^{2k-1})$. There are 5^{k-1} values for r_2 in Lemma 5.33, and for each value of r_2 there are 5^{k-1} elements of the complete set of residues divisible by $2 + \omega$. Consequently, since there are $|N((2 + \omega)^{2k-1})| = 5^{2k-1}$ elements in the complete set of residues, then

$$\Phi(\pi^{2k-1}) = 5^{2k-1} - (5^{k-1})(5^{k-1}) = 5^{2k-1}(1 - 1/5)$$

$$= |N(\pi^{2k-1})| \cdot \left(1 - |N(\pi)|^{-1}\right). \quad \text{Q.E.D.}$$

LEMMA 5.35 *If $k > 0$, then the integers in the set $\{r_1 + r_2\omega : 0 \le r_1 < 5^k, 0 \le r_2 < 5^k\}$ of complete residues modulo $(2 + \omega)^{2k}$ that are divisible by $2 + \omega$ are the integers given in (i)–(v) of Lemma 5.33.*
PROOF: An easy consequence of Lemma 5.32. Q.E.D.

LEMMA 5.36 *If $k > 0$ and $|N(\pi)| = 5$, then*
$$\Phi(\pi^{2k}) = |N(\pi^{2k})| \cdot \left(1 - |N(\pi)|^{-1}\right).$$

PROOF: Use Lemma 5.35 and proceed as in Lemma 5.34. Q.E.D.

LEMMA 5.37 *If $k > 0$ and π is an associate of a rational prime p of the form $p \equiv \pm 2 \pmod 5$, then $\Phi(\pi^k)$ is given by (5.2).*

PROOF: We have $\Phi(\pi^k) = \Phi(p^k)$. The integers in the set $\{r_1 + r_2\omega : 0 \le r_1 < p^k, 0 \le r_2 < p^k\}$ of complete residues modulo p^k that are divisible by p are $r_1 + r_2\omega$, where r_1 and r_2 are from the set $\{0, p, 2p, 3p, 4p, \ldots, (p^{k-1} - 1)p\}$. There are $(p^{k-1})(p^{k-1}) = p^{2k-2}$ such integers. Consequently,

$$\Phi(\pi^k) = |N(p^k)| - p^{2k-2} = p^{2k} - p^{2k-2}$$
$$= p^{2k}(1 - p^{-2})$$
$$= |N(\pi^k)| \cdot (1 - |N(\pi)|^{-1}). \quad \text{Q.E.D.}$$

LEMMA 5.38 *If $k > 0$ and π is a prime such that $|N(\pi)| = q$, where q is a rational prime of the form $q \equiv \pm 1 \pmod 5$, then $\Phi(\pi^k)$ is given by (5.2).*

PROOF: The integers in the set $\{0, 1, 2, \ldots, q^k - 1\}$ of complete residues modulo π^k that are divisible by π are $0, q, 2q, 3q, \ldots, (q^{k-1} - 1)q$. There are q^{k-1} such integers. Consequently,

$$\Phi(\pi^k) = |N(\pi^k)| - q^{k-1} = q^k - q^{k-1}$$
$$= q^k(1 - q^{-1})$$
$$= |N(\pi^k)| \cdot (1 - |N(\pi)|^{-1}). \quad \text{Q.E.D.}$$

THEOREM 5.39 *If $k > 0$ and π is a prime, then $\Phi(\pi^k)$ is given by (5.2).*

PROOF: The result follows from Lemmas 5.34, 5.36, 5.37, and 5.38. Q.E.D.

Chinese Remainder Theorem for $\hat{Z}$

THEOREM 5.40 (Chinese remainder theorem for $\hat{Z}$) *If the non-zero integers μ_i, $i = 1, 2, \ldots, k$, are relatively prime in pairs, and if*

α_i, $i = 1, 2, \ldots, k$, *are any given integers, then the system of congruences*

$$\xi \equiv \alpha_1 \ (\text{MOD } \mu_1), \qquad \xi \equiv \alpha_2 \ (\text{MOD } \mu_2) \quad, \ldots, \qquad \xi \equiv \alpha_k \ (\text{MOD } \mu_k)$$

admits a simultaneous solution. Furthermore, this solution is unique modulo $\mu = \mu_1 \mu_2 \cdots \mu_k$.

PROOF: Let $\mu = \mu_1 \mu_2 \cdots \mu_k$ and write $\mu = \mu_i M_i$ for $i = 1, 2, \ldots, k$. Then $\text{GCD}(\mu_i, M_i) = 1$ for each i, so that for each i there exist η_i and ζ_i such that $M_i \eta_i + \mu_i \zeta_i = 1$. Consequently, $M_i \eta_i \equiv 1$ (MOD μ_i) and $M_i \eta_i \equiv 0$ (MOD M_i) for each i. From the last congruence we deduce that $M_i \eta_i \equiv 0$ (MOD μ_j) for each $j \neq i$. Now consider the integer ξ given by

$$\xi = \sum_{i=1}^{k} \alpha_i M_i \eta_i.$$

For $j = 1, 2, \ldots, k$ we have

$$\xi = \sum_{i=1}^{k} \alpha_i M_i \eta_i \equiv \alpha_j M_j \eta_j \equiv \alpha_j \ (\text{MOD } \mu_j).$$

Thus ξ is a simultaneous solution of the system of k congruences.

Also, if η is any other simultaneous solution, then $\eta \equiv \alpha_i \equiv \xi$ (MOD μ_i) for $i = 1, 2, \ldots, k$. Since the μ_i are relatively prime in pairs, this entails that $\eta \equiv \xi$ (MOD $\mu_1 \mu_2 \cdots \mu_k$). Hence the solution ξ is unique modulo $\mu_1 \mu_2 \cdots \mu_k$. Q.E.D.

It should be noted that the preceding proof gives a practical method whereby the simultaneous solution can be computed.

Verification of Formula (5.1) and Further Results on the Totient

THEOREM 5.41 *If* $\text{GCD}(\mu_1, \mu_2) = 1$, *then* $\Phi(\mu_1 \mu_2) = \Phi(\mu_1) \cdot \Phi(\mu_2)$.

PROOF: Let $\{\eta_1, \eta_2, \ldots, \eta_j\}$ and $\{\zeta_1, \zeta_2, \ldots, \zeta_k\}$ be reduced sets of residues modulo μ_1 and μ_2, respectively. If ξ is in a reduced residue set modulo $\mu_1\mu_2$, then $\text{GCD}(\xi, \mu_1) = \text{GCD}(\xi, \mu_2) = 1$, so that $\xi \equiv \eta_h$ (MOD μ_1) and $\xi = \zeta_i$ (MOD μ_2) for some h and i. Conversely, if $\xi \equiv \eta_h$ (MOD μ_1) and $\xi \equiv \zeta_i$ (MOD μ_2), then $\text{GCD}(\xi, \mu_1) = \text{GCD}(\xi, \mu_2) = 1$, so that $\text{GCD}(\xi, \mu_1\mu_2) = 1$. Thus a reduced set of residues modulo $\mu_1\mu_2$ can be obtained by determining all ξ such that $\xi \equiv \eta_h$ (MOD μ_1) and $\xi \equiv \zeta_i$ (MOD μ_2) for some h and i. According to the Chinese remainder theorem, each pair (h, i) determines a single ξ modulo $\mu_1\mu_2$; and different pairs (h, i) yield different ξ modulo $\mu_1\mu_2$. There are $jk = \Phi(\mu_1) \cdot \Phi(\mu_2)$ of these pairs. Therefore a set of reduced residues modulo $\mu_1\mu_2$ contains $jk = \Phi(\mu_1) \cdot \Phi(\mu_2)$ elements, and we have $\Phi(\mu_1\mu_2) = \Phi(\mu_1) \cdot \Phi(\mu_2)$. Q.E.D.

Theorem 5.41 has completed the necessary preliminaries to establish

THEOREM 5.42 *If μ is not a unit and $\pi_1, \pi_2, \ldots, \pi_n$ are the nonassociated prime factors of μ, then $\Phi(\mu)$ is given by (5.1).*
PROOF: We have $\mu = \epsilon \cdot \pi_1^{k_1}\pi_2^{k_2} \cdots \pi_n^{k_n}$, where ϵ is a unit, and the k_i are positive. By Theorems 5.39 and 5.41,

$$\Phi(\mu) = \Phi(\epsilon) \cdot \Phi(\pi_1^{k_1}) \cdots \Phi(\pi_n^{k_n})$$

$$= |N(\pi_1^{k_1}\pi_2^{k_2} \cdots \pi_n^{k_n})| \cdot \prod_{i=1}^{n} \left(1 - |N(\pi_i)|^{-1}\right)$$

$$= |N(\mu)| \cdot \prod_{i=1}^{n} \left(1 - |N(\pi_i)|^{-1}\right). \quad \text{Q.ED.}$$

EXAMPLE 5.5 From Example 3.3 we have

$$-1152 + 1324\omega = \omega^{-3}(2 + \omega)(2^2)(5 + \omega)^3.$$

Since $N(2 + \omega) = 5$, $N(2) = 4$, and $N(5 + \omega) = 29$, we obtain

$$\Phi(-1152 + 1324\omega) = (5)(16)(29^3)(1 - 5^{-1})(1 - 4^{-1})(1 - 29^{-1})$$

$$= (4)(29^2)(4)(3)(28)$$

$$= 1130304.$$

The following lemma will allow us in Theorem 5.44 and Corollary 5.45 to compare the values of $\Phi(n)$ and $\phi(n)$ when $n > 1$.

LEMMA 5.43 *Let p be a rational prime and* $k > 0$.
(i) *If* $p \equiv \pm 2 \pmod 5$, *then*

$$\Phi(p^k) = p^{2k-2}(p+1)(p-1).$$

(ii) *If* $p \equiv \pm 1 \pmod 5$, *then*

$$\Phi(p^k) = p^{2k-2}(p-1)^2.$$

(iii) $\Phi(5^k) = 5^{2k-2}(5)(5-1)$.
PROOF: If $p \equiv \pm 2 \pmod 5$, then p is a prime. Therefore $\Phi(p^k)$ $= p^{2k}(1 - p^{-2}) = p^{2k-2}(p+1)(p-1)$. If $p \equiv \pm 1 \pmod 5$, then $p = N(\pi) = \pi\bar{\pi}$, where π and $\bar{\pi}$ are nonassociated primes. Consequently, $\Phi(p^k) = \Phi(\pi^k) \cdot \Phi(\bar{\pi}^k) = p^{2k}(1 - 1/p)^2 = p^{2k-2}(p-1)^2$. Since $(2 + \omega)^2 = 5\omega^2$, then 5^k and $(2 + \omega)^{2k}$ are associates. Thus $\Phi(5^k) = \Phi((2 + \omega)^{2k}) = 5^{2k}(1 - 1/5) = 5^{2k-2}(5)(5-1)$. Q.E.D.

THEOREM 5.44 *If* $n > 1$, *then* $\Phi(n) \geq (\phi(n))^2$.
PROOF: Let $n = p_1^{k_1} p_2^{k_2} \cdots p_r^{k_r}$, where the p_i are distinct rational primes and the k_i are positive. Since the $p_i^{k_i}$ are relatively prime in pairs, then

$$\Phi(n) = \Phi(p_1^{k_1}) \cdot \Phi(p_2^{k_2}) \cdots \Phi(p_r^{k_r})$$

and

$$\phi(n) = \phi(p_1^{k_1}) \cdot \phi(p_2^{k_2}) \cdots \phi(p_r^{k_r}).$$

Thus it suffices to show that $\Phi(p^k) \geq (\phi(p^k))^2$ for each rational prime p and $k > 0$. Since $\phi(p^k) = p^{k-1}(p-1)$, then $(\phi(p^k))^2$ $= p^{2k-2}(p-1)^2$. The desired result is now evident from Lemma 5.43. Q.E.D.

COROLLARY 5.45 *If* $n > 1$, *then* $\Phi(n) > \phi(n)$.
PROOF: If $n > 2$, then $\phi(n) > 1$, and so $\Phi(n) \geq (\phi(n))^2 > \phi(n)$. Also, $\phi(2) = 1$ and $\Phi(2) = |N(2)| - 1 = 3$. Thus the desired inequality holds for all $n > 1$. Q.E.D.

Two Theorems on Reduced Residue Sets

Analogous to Theorems 5.18 and 5.19 for complete residue sets, we have the following two results for reduced residue sets.

THEOREM 5.46 *If* $\{\xi_1, \xi_2, \ldots, \xi_t\}$ *is a reduced set of residues modulo* μ *and* $GCD(\alpha, \mu) = 1$, *then* $\{\alpha\xi_1, \alpha\xi_2, \ldots, \alpha\xi_t\}$ *is a reduced set of residues modulo* μ.

PROOF: Since $GCD(\xi_i, \mu) = 1$ and $GCD(\alpha, \mu) = 1$, then $GCD(\alpha\xi_i, \mu) = 1$ for $i = 1, 2, \ldots, t$. Also $\alpha\xi_i \equiv \alpha\xi_j$ (MOD μ) if and only if $\xi_i \equiv \xi_j$ (MOD μ). Hence the $t = \Phi(\mu)$ integers $\alpha\xi_i$ are relatively prime to μ and incongruent modulo μ, and thus form a reduced set of residues modulo μ. Q.E.D.

THEOREM 5.47 *If* $GCD(\mu_1, \mu_2) = 1$, $\{\xi_1, \xi_2, \ldots, \xi_t\}$ *is a reduced set of residues modulo* μ_1, *and* $\{\eta_1, \eta_2, \ldots, \eta_s\}$ *is a reduced set of residues modulo* μ_2, *then the set* $\{\mu_2\xi_i + \mu_1\eta_j : i = 1, 2, \ldots, t, j = 1, 2, \ldots, s\}$ *is a reduced set of residues modulo* $\mu_1\mu_2$.

PROOF: We have $GCD(\mu_2\xi_i + \mu_1\eta_j, \mu_1) = GCD(\mu_2\xi_i, \mu_1) = 1$, and $GCD(\mu_2\xi_i + \mu_1\eta_j, \mu_2) = GCD(\mu_1\eta_j, \mu_2) = 1$ for each i and j. Just as in the proof of Theorem 5.19, we conclude that the st elements $\mu_2\xi_i + \mu_1\eta_j$ are incongruent modulo $\mu_1\mu_2$. Hence the $\Phi(\mu_1\mu_2) = \Phi(\mu_1) \cdot \Phi(\mu_2) = st$ integers $\mu_2\xi_i + \mu_1\eta_j$ are relatively prime to $\mu_1\mu_2$ and incongruent modulo $\mu_1\mu_2$, and thus form a reduced set of residues modulo $\mu_1\mu_2$. Q.E.D.

EXAMPLE 5.6 $\{1, 2, 3, 4\}$ is a reduced set of residues modulo $1 + 3\omega$, and $\{1, \omega, 1 + \omega\}$ is a reduced set of residues modulo 2. Therefore, by Theorem 5.47, we conclude that $\{(1 + 3\omega)(r_1 + r_2\omega) + 2k : 0 \leq r_1 \leq 1, 0 \leq r_2 \leq 1, r_1 + r_2 \neq 0, 1 \leq k \leq 4\}$ is a reduced set of residues modulo $2 + 6\omega$. Listing these twelve elements, we obtain the set $\{9 + 4\omega, 11 + 4\omega, 5 + 4\omega, 7 + 4\omega, 12 + 7\omega, 5 + 3\omega, 7 + 3\omega, 6 + 7\omega, 9 + 3\omega, 8 + 7\omega, 10 + 7\omega, 3 + 3\omega\}$. (Compare with Example 5.4.)

A Number-theoretic Proof of the Euler–Fermat Theorem

Using Theorem 5.46, we have the following number-theoretic proof of the Euler–Fermat theorem. Both $\{\xi_1, \xi_2, \ldots, \xi_t\}$ and $\{\alpha\xi_1, \alpha\xi_2, \ldots, \alpha\xi_t\}$ are reduced residue sets modulo μ, and so $\alpha\xi_1$, $\alpha\xi_2, \ldots, \alpha\xi_t$ are congruent modulo μ to $\xi_1, \xi_2, \ldots, \xi_t$, but not necessarily in the same order. Thus, by multiplication of congruences,

$$\alpha\xi_1 \alpha\xi_2 \cdots \alpha\xi_t \equiv \xi_1 \xi_2 \cdots \xi_t \ (\text{MOD } \mu).$$

Since $t = \Phi(\mu)$, this last congruence gives

$$\alpha^{\Phi(\mu)} \xi_1 \xi_2 \cdots \xi_t \equiv \xi_1 \xi_2 \cdots \xi_t \ (\text{MOD } \mu).$$

But $\text{GCD}(\xi_1 \xi_2 \cdots \xi_t, \mu) = 1$ and, by cancellation, we obtain $\alpha^{\Phi(\mu)} \equiv 1 \ (\text{MOD } \mu)$.

A Summation Formula for Φ

In this section we obtain the analogue of the summation formula

$$\sum_{d \mid n} \phi(d) = n,$$

where the sum is extended over all $d > 0$ such that d divides n $(n > 0)$.

LEMMA 5.48 If $\alpha \mid \mu$ and $\{\xi_1, \xi_2, \ldots, \xi_t\}$ is a complete set of residues modulo μ, then exactly $\Phi(\mu/\alpha)$ of the ξ_i satisfy the condition $\text{GCD}(\xi_i, \mu) = \alpha$.

PROOF: Let $\{\eta_1, \eta_2, \ldots, \eta_s\}$ be a reduced set of residues modulo

μ/α, so that $s = \Phi(\mu/\alpha)$ and $\mathrm{GCD}(\eta_i, \mu/\alpha) = 1$ for $i = 1$, $2, \ldots, s$. Consider the set $\{\alpha\eta_1, \alpha\eta_2, \ldots, \alpha\eta_s\}$. Since $\alpha \mid \mu$, then $\alpha = \mathrm{GCD}(\alpha, \mu)$, so that $\alpha\eta_i \equiv \alpha\eta_j$ (MOD μ) if and only if $\eta_i \equiv \eta_j$ (MOD μ/α). Thus the s elements $\alpha\eta_i$ are incongruent modulo μ and are therefore congruent modulo μ to s distinct elements in the set $\{\xi_1, \xi_2, \ldots, \xi_t\}$. By reordering the subscripts, we may assume that $\xi_j \equiv \alpha\eta_j$ (MOD μ) for $j = 1, 2, \ldots, s$. As the elements ξ_j and $\alpha\eta_j$ are in the same residue class modulo μ, then $\mathrm{GCD}(\xi_j, \mu)$ $= \mathrm{GCD}(\alpha\eta_j, \mu)$. Consequently, since $\mathrm{GCD}(\eta_j, \mu/\alpha) = 1$, then $\mathrm{GCD}(\xi_j, \mu) = \mathrm{GCD}(\alpha\eta_j, \mu) = \alpha \cdot \mathrm{GCD}(\eta_j, \mu/\alpha) = \alpha$ for $j = 1$, $2, \ldots, s$. At this point we have found $s = \Phi(\mu/\alpha)$ of the ξ_i that satisfy the condition $\mathrm{GCD}(\xi_i, \mu) = \alpha$; namely, $\xi_1, \xi_2, \ldots, \xi_s$. Suppose, to the contrary, that there exists an $i > s$ such that $\mathrm{GCD}(\xi_i, \mu) = \alpha$. Then $\xi_i = \alpha\delta$ for some δ, and $\mathrm{GCD}(\delta, \mu/\alpha) = \mathrm{GCD}(\xi_i/\alpha, \mu/\alpha) = 1$. Therefore $\delta \equiv \eta_j$ (MOD μ/α) for some $j \leq s$. Since $\alpha = \mathrm{GCD}(\alpha, \mu)$ and $\delta \equiv \eta_j$ (MOD μ/α), then $\alpha\delta \equiv \alpha\eta_j$ (MOD μ). Accordingly, $\xi_i = \alpha\delta \equiv \alpha\eta_j \equiv \xi_j$ (MOD μ), where $i > j$. This contradicts the fact that $\xi_i \not\equiv \xi_j$ (MOD μ) when $i \neq j$. Thus there does not exist an $i > s$ such that $\mathrm{GCD}(\xi_i, \mu) = \alpha$, and the proof is complete. Q.E.D.

THEOREM 5.49 (Summation formula for Φ) *If $\mu \neq 0$, then*

$$\sum_{\delta \mid \mu} \Phi(\delta) = |\mathrm{N}(\mu)|,$$

where the sum is extended over any set of $\hat{\tau}(\mu)$ nonassociated divisors of μ.

PROOF: Let $\{\delta_1, \delta_2, \ldots, \delta_n\}$ be a set of $n = \hat{\tau}(\mu)$ nonassociated divisors of μ. Then $\mu = \delta_1\gamma_1 = \delta_2\gamma_2 = \ldots = \delta_n\gamma_n$, and it is clear that $\{\gamma_1, \gamma_2, \ldots, \gamma_n\}$ is also a set of nonassociated divisors of μ. Let $\{\xi_1, \xi_2, \ldots, \xi_t\}$, where $t = |\mathrm{N}(\mu)|$, be a complete set of residues modulo μ. We partition the integers of the set $\{\xi_1, \xi_2, \ldots, \xi_t\}$ into n disjoint subsets as follows: Place in the first subset those ξ_i such that $\mathrm{GCD}(\xi_i, \mu) = \gamma_1$. By Lemma 5.48, there are exactly $\Phi(\mu/\gamma_1)$ $= \Phi(\delta_1)$ of the ξ_i which are in this subset. Place in the second subset those ξ_i such that $\mathrm{GCD}(\xi_i, \mu) = \gamma_2$. There are exactly $\Phi(\mu/\gamma_2) = \Phi(\delta_2)$ of the ξ_i which are in this subset. Proceeding in this way, it is evident that we shall have n subsets; and since every ξ_i has some divisor of μ as its greatest common divisor with μ, then each ξ_i in the set $\{\xi_1, \xi_2, \ldots, \xi_t\}$ will occur in exactly one of the n subsets. Also, the ith subset will have $\Phi(\mu/\gamma_i) = \Phi(\delta_i)$ elements. Hence $|N(\mu)| = t = \Phi(\delta_1) + \Phi(\delta_2) + \ldots + \Phi(\delta_n)$, which is the assertion of the theorem. Q.E.D.

EXAMPLE 5.7 A set of $6 = \hat{\tau}(\mu)$ nonassociated divisors of $\mu = 3^2(3 + \omega)$ is $\{1, 3, 3^2, 3 + \omega, 3(3 + \omega), 3^2(3 + \omega)\}$. We have $\Phi(1) + \Phi(3) + \Phi(3^2) + \Phi(3 + \omega) + \Phi(3(3 + \omega)) + \Phi(3^2(3 + \omega)) = 1 + 8 + 72 + 10 + 80 + 720 = 891 = |N(\mu)|$.

An Important Generalization of Fermat's Theorem

Fermat's theorem for rational integers states that if p is a rational prime, then $a^p \equiv a \pmod{p}$ for any a. As previously noted, the last congruence is equivalent to $a^p \equiv a \pmod{p}$. Theorem 5.29 generalized this theorem to $\hat{Z}$. We now develop another generalization of this important theorem. We shall show that if $p \neq 5$ is a rational prime and α is any integer, then $\alpha^p \equiv \bar{\alpha}$ (MOD p) when $p \equiv \pm 2 \pmod{5}$, and $\alpha^p \equiv \alpha$ (MOD p) when $p \equiv \pm 1 \pmod{5}$. Note that if α is a rational integer, then the preceding statement, apart from the case $p = 5$, is a confirmation of Fermat's theorem for rational integers. A preliminary lemma is first established, the proof of which relies heavily on the quadratic reciprocity law and Euler's criterion for quadratic residues.

LEMMA 5.50 *If p is a rational prime of the form $p \equiv \pm 2 \pmod{5}$, then $\omega^p \equiv \bar{\omega} \pmod{p}$.*
PROOF: Since $\omega^2 - \bar{\omega} = \omega + 1 - (1 - \omega) = 2\omega$, then $2 \mid (\omega^2 - \bar{\omega})$; and so $\omega^2 \equiv \bar{\omega} \pmod{2}$. Thus the lemma is true for $p = 2$.

Now suppose $p \equiv \pm 2 \pmod{5}$ and $p \neq 2$, so that $\gcd(2, p) = 1$. Since $2^p \equiv 2 \pmod{p}$, by Fermat's theorem for rational integers, then we have

$$2\omega^p \equiv (2\omega)^p = \left(1 + \sqrt{5}\,\right)^p \equiv 1 + 5^{(p-1)/2}\sqrt{5} \quad (\text{MOD } p). \quad (5.3)$$

By the quadratic reciprocity law for rational integers, we have $(5|p) = (p|5)$; and since $p \equiv \pm 2$ (mod 5), then p is not a quadratic residue modulo 5. Thus $(5|p) = (p|5) = -1$. We therefore have $5^{(p-1)/2} \equiv -1$ (MOD p), by Euler's criterion for quadratic residues. Thus it follows from (5.3) that $2\omega^p \equiv 1 - \sqrt{5} = 2\overline{\omega}$ (MOD p); and since $\gcd(2, p) = 1$, we obtain $\omega^p \equiv \overline{\omega}$ (MOD p), by cancellation. Q.E.D.

THEOREM 5.51 *If $p \neq 5$ is a rational prime and α is any integer, then*
(i) $\alpha^p \equiv \overline{\alpha}$ (MOD p), *when* $p \equiv \pm 2$ (mod 5);
(ii) $\alpha^p \equiv \alpha$ (MOD p), *when* $p \equiv \pm 1$ (mod 5).

PROOF: Let $\alpha = c + d\omega$. By Fermat's theorem for rational integers, $c^p \equiv c$ (MOD p) and $d^p \equiv d$ (MOD p). Furthermore, if $p \equiv \pm 2$ (mod 5), then $\omega^p \equiv \overline{\omega}$ (MOD p), by Lemma 5.50. Accordingly,

$$\alpha^p = (c + d\omega)^p \equiv c^p + d^p\omega^p \equiv c + d\overline{\omega} = \overline{\alpha} \quad (\text{MOD } p)$$

when $p \equiv \pm 2$ (mod 5).

On the other hand, if $p \equiv \pm 1$ (mod 5), then we have $p = \pi\overline{\pi}$, where π and $\overline{\pi}$ are nonassociated primes. Thus $\alpha^p \equiv \alpha$ (MOD π) and $\alpha^p \equiv \alpha$ (MOD $\overline{\pi}$), by Theorem 5.29(i). Since $\text{GCD}(\pi, \overline{\pi}) = 1$, we then have $\alpha^p \equiv \alpha$ (MOD $\pi\overline{\pi}$), or, equivalently, $\alpha^p \equiv \alpha$ (MOD p). Q.E.D.

Addendum to Example 2.3

We conclude the chapter by addressing the comment following Example 2.3: There are an infinite number of ρ such that $|\text{N}(\rho)| < 5$ and $2 = (-1 + 2\omega)\theta + \rho$ for some θ depending on ρ. In order to verify the foregoing statement, we cast it in the equivalent form: There are an infinite number of ρ such that $|\text{N}(\rho)| < 5$ and $\rho \equiv 2$ (MOD $-1 + 2\omega$).

Let us suppose that $\rho \equiv 2$ (MOD $-1 + 2\omega$) and $|N(\rho)| < 5$. Since 2 and 3 are primes (of $\hat{Z}$), then there does not exist a ρ such that $|N(\rho)|$ is 2 or 3. Also, since $\rho \neq 0$, then $|N(\rho)| \neq 0$. Consequently, $|N(\rho)|$ is 1 or 4, so that $\rho = \epsilon$ or $\rho = 2\epsilon$ for some unit ϵ. It is easily verified that, modulo $-1 + 2\omega$, we have

$$\omega^3 \equiv 2, \qquad -\omega \equiv 2,$$
$$2 \equiv 2, \qquad -2\omega^2 \equiv 2.$$

Also, $\Phi(-1 + 2\omega) = 4$, so that, by the Euler–Fermat theorem, $\omega^{4n} \equiv 1$ (MOD $-1 + 2\omega$) for any n (positive, negative, or zero). Thus, modulo $-1 + 2\omega$, we have

$$\omega^{4n+3} \equiv 2, \qquad -\omega^{4n+1} \equiv 2,$$
$$2\omega^{4n} \equiv 2, \qquad -2\omega^{4n+2} \equiv 2,$$

and ρ may be taken to be any of the integers on the left sides of the preceding four congruences. A little more work using some of the results of Chapter 7 (in particular, Theorem 7.2(iii)) would reveal that there are no other values for ρ.

VI   Algebraic Congruences

Introduction

In this chapter we discuss the solution of the algebraic congruence $f(X) \equiv 0$ (MOD μ), where $f(X)$ is a polynomial with integer coefficients (coefficients in $\hat{Z}$). The simplest such congruence is the linear congruence $\alpha X \equiv \beta$ (MOD μ). After dispensing with the linear congruence, we proceed to the general algebraic congruence. We begin by restricting μ to be a prime π. Next we allow μ to be a power of a prime π; and we develop an inductive procedure whereby the solutions of $f(X) \equiv 0$ (MOD π^k) are determined from the solutions of $f(X) \equiv 0$ (MOD π^{k-1}). Finally, if $\mu = \pi_1^{c_1} \pi_2^{c_2} \cdots \pi_r^{c_r}$, where the π_i are nonassociated primes and $c_i > 0$, then a procedure based on the Chinese remainder theorem will be developed whereby the solutions of $f(X) \equiv 0$ (MOD μ) are determined from the solutions of $f(X) \equiv 0$ (MOD $\pi_i^{c_i}$), $i = 1, 2, \ldots, r$.

Thus, if $\mu = \pi_1^{c_1} \pi_2^{c_2} \cdots \pi_r^{c_r}$ and the solutions of the algebraic congruences $f(X) \equiv 0$ (MOD π_i), $i = 1, 2, \ldots, r$, are known, then we will be able to employ the procedures developed in this chapter to find the solutions of $f(X) \equiv 0$ (MOD μ). Unfortunately, there is no general method for solving $f(X) \equiv 0$ (MOD π_i), although special techniques do exist when $f(X)$ is of a special nature. (The linear congruence is such a special case.) For the general congruence $f(X) \equiv 0$ (MOD π_i), direct substitution of the integers in a complete set of residues modulo π_i is often the best method available for solving the congruence.

Polynomials over $\hat{Z}_\mu$

THEOREM 6.1   *If* $F(X) = \alpha_0{}^* + \alpha_1{}^*X + \ldots + \alpha_n{}^*X^n$ *is a polynomial over* $\hat{Z}_\mu$, *then* ξ^* *is a root of* $F(X)$ *if and only if* $\alpha_0 + \alpha_1\xi + \ldots + \alpha_n\xi^n \equiv 0$ (MOD μ).
PROOF: The equation $\alpha_0{}^* + \alpha_1{}^*\xi^* + \ldots + \alpha_n{}^*(\xi^*)^n = 0^*$ is equivalent to $(\alpha_0 + \alpha_1\xi + \ldots + \alpha_n\xi^n)^* = 0^*$, and ξ satisfies the last equation if and only if $\alpha_0 + \alpha_1\xi + \ldots + \alpha_n\xi^n \equiv 0$ (MOD μ). Q.E.D.

DEFINITIONS   If

$$f(X) = \alpha_0 + \alpha_1 X + \ldots + \alpha_n X^n \tag{6.1}$$

then the expression

$$f(X) \equiv 0 \text{ (MOD } \mu) \tag{6.2}$$

is called an **algebraic congruence modulo μ in the indeterminate X**, and μ is called the **modulus** of the algebraic congruence. An integer ξ is said to be a **solution** of (6.2) if $f(\xi) \equiv 0$ (MOD μ). The algebraic congruence (6.2) is said to be an **nth-degree** algebraic congruence if $\alpha_n \not\equiv 0$ (MOD μ). The polynomial

$$F(X) = \alpha_0{}^* + \alpha_1{}^*X + \ldots + \alpha_n{}^*X^n \tag{6.3}$$

is called the **polynomial over $\hat{Z}_\mu$ corresponding to the polynomial (6.1).**

The following three corollaries are immediate consequences of Theorem 6.1 and the above definitions.

COROLLARY 6.2   *The integer ξ is a solution of (6.2) if and only if ξ^* is a root of the polynomial (6.3).*

COROLLARY 6.3   *If ξ is a solution of (6.2) and $\eta \equiv \xi$ (MOD μ), then η is also a solution of (6.2).*

COROLLARY 6.4   *The number of incongruent solutions modulo μ of (6.2) is equal to the number of distinct roots of the polynomial (6.3). Moreover, $\xi_1, \xi_2, \ldots, \xi_k$ are the incongruent solutions modulo μ of (6.2) if and only if $\xi_1^*, \xi_2^*, \ldots, \xi_k^*$ are the distinct roots of the polynomial (6.3).*

Linear Congruences

THEOREM 6.5   *If GCD$(\alpha, \mu) = 1$, then the linear congruence $\alpha X \equiv \beta$ (MOD μ) has a unique solution modulo μ.*
PROOF: Since GCD$(\alpha, \mu) = 1$, there exist η and ζ such that $1 = \alpha\eta + \mu\zeta$. Thus $\alpha\eta \equiv 1$ (MOD μ), and so $\alpha(\eta\beta) \equiv \beta$ (MOD μ). Hence $\xi = \eta\beta$ is a solution of the linear congruence.

Suppose now that ξ_1 and ξ_2 are both solutions of the linear congruence. Then $\alpha\xi_1 \equiv \beta = \alpha\xi_2$ (MOD μ), and since GCD$(\alpha, \mu) = 1$, we obtain $\xi_1 \equiv \xi_2$ (MOD μ). Consequently, all solutions of the linear congruence are congruent modulo μ. Q.E.D.

LEMMA 6.6   *IF GCD$(\alpha, \mu) = \delta$, $\alpha = \alpha_1\delta$, $\mu = \mu_1\delta$, and $\beta = \beta_1\delta$, then the linear congruence*

$$\alpha X \equiv \beta \text{ (MOD } \mu) \tag{6.4}$$

has a solution. Furthermore, ξ is a solution of (6.4) if and only if ξ is a solution of

$$\alpha_1 X \equiv \beta_1 \text{ (MOD } \mu_1). \tag{6.5}$$

PROOF: Suppose that ξ is a solution of (6.5). Then $\mu_1 | (\alpha_1\xi - \beta_1)$, so that $\mu_1\delta | (\alpha_1\delta\xi - \beta_1\delta)$. Accordingly, $\mu | (\alpha\xi - \beta)$, and so ξ is a solution of (6.4). Conversely, suppose that ξ is a solution of (6.4). Then $\mu_1\delta \,|\, (\alpha_1\delta\xi - \beta_1\delta)$, so that $\mu_1 \,|\, (\alpha_1\xi - \beta_1)$. Hence ξ is a solution of (6.5).

Since $\mathrm{GCD}(\alpha_1, \mu_1) = 1$, it follows from Theorem 6.5 that (6.5) has a solution ξ. Consequently, by what was established in the last paragraph, ξ is a solution of (6.4). Q.E.D.

THEOREM 6.7 *IF* $\mathrm{GCD}(\alpha, \mu) = \delta$, *then the linear congruence* (6.4) *has a solution if and only if* $\delta \,|\, \beta$. *If* $\delta \,|\, \beta$, $\{\zeta_1, \zeta_2, \ldots, \zeta_r\}$ *is a complete set of residues modulo* δ, $\mu = \mu_1\delta$, *and* ξ *is any solution of the linear congruence* (6.4), *then the* r *integers* $\xi + \mu_1\zeta_i$, $i = 1$, $2, \ldots, r$, *are the incongruent solutions modulo* μ *of* (6.4). *Thus, if* $\delta \,|\, \beta$, *then* (6.4) *has exactly* $|\mathrm{N}(\delta)|$ *incongruent solutions modulo* μ.

PROOF: If $\delta \,|\, \beta$, then it follows from Lemma 6.6 that (6.4) has a solution. Conversely, suppose that (6.4) has a solution, so that $\alpha\xi \equiv \beta$ (MOD μ) for some ξ. Since $\delta \,|\, \alpha$ and $\delta \,|\, \mu$, then $\alpha \equiv 0$ (MOD δ) and $\alpha\xi \equiv \beta$ (MOD δ). Therefore $\beta \equiv \alpha\xi \equiv 0\xi = 0$ (MOD δ), and so $\delta \,|\, \beta$.

If $\xi + \mu_1\zeta_i \equiv \xi + \mu_1\zeta_j$ (MOD μ), then $\mu_1\zeta_i \equiv \mu_1\zeta_j$ (MOD μ); and since $\mathrm{GCD}(\mu_1, \mu) = \mu_1$ and $\delta = \mu/\mu_1$, then $\zeta_i \equiv \zeta_j$ (MOD δ). Thus $i = j$ and we conclude that the r integers $\xi + \mu_1\zeta_i$ are incongruent modulo μ. Using the notation of Lemma 6.6, we have $\alpha(\xi + \mu_1\zeta_i) = \alpha\xi + \alpha\mu_1\zeta_i = \alpha\xi + \alpha_1\mu_1\delta\zeta_i = \alpha\xi + \alpha_1\zeta_i\mu \equiv \alpha\xi \equiv \beta$ (MOD μ). Thus we have shown that the r integers $\xi + \mu_1\zeta_i$ are incongruent solutions modulo μ of (6.4).

Finally, suppose that η is a solution of (6.4). We must establish that η is congruent modulo μ to one of the r integers $\xi + \mu_1\zeta_i$. It follows from Lemma 6.6 that both η and ξ are solutions of (6.5). Also, since $\mathrm{GCD}(\alpha_1, \mu_1) = 1$, we conclude from Theorem 6.5 that $\eta \equiv \xi$ (MOD μ_1). Hence $\eta = \xi + \mu_1\gamma$, and since there exists a ζ_i such that $\gamma = \zeta_i + \theta\delta$, then $\eta = \xi + (\zeta_i + \theta\delta)\mu_1 = \xi + \mu_1\zeta_i + \theta\mu_1\delta = \xi + \mu_1\zeta_i + \theta\mu \equiv \xi + \mu_1\zeta_i$ (MOD μ). Therefore η is congruent modulo μ to one of the r integers $\xi + \mu_1\zeta_i$. Q.E.D.

EXAMPLE 6.1 Since $\mathrm{GCD}(4, 6 + 2\omega) = 2 \cdot \mathrm{GCD}(2, 3 + \omega) = 2$ and $2 \,|\, 6$, then the linear congruence $4X \equiv 6$ (MOD $6 + 2\omega$) has $|\mathrm{N}(2)| = 4$ incongruent solutions modulo $6 + 2\omega$. From Lemma 6.6 we deduce that a solution of $2X \equiv 3$ (MOD $3 + \omega$) is also a solution of $4X \equiv 6$ (MOD $6 + 2\omega$). By inspection, we observe that $(3 + \omega)(1) + (-1)(2) = 1 + \omega = \omega^2$ and, as $\omega^{-2} = F_3 - F_2\omega = 2 - \omega$, then $(3 + \omega)(2 - \omega) + (-2 + \omega)(2) = 1$. Thus $2(-2 + \omega) \equiv 1$ (MOD $3 + \omega$). Consequently, $-6 + 3\omega = 3(-2 + \omega)$ is a solution of $2X \equiv 3$ (MOD $3 + \omega$), and therefore a solution of $4X \equiv 6$ (MOD $6 + 2\omega$). Also, $\{0, 1, \omega, 1 + \omega\}$ is a complete set of residues modulo 2. Accordingly, it follows from Theorem 6.7 that the four incongruent solutions modulo $6 + 2\omega$ of $4X \equiv 6$ (MOD $6 + 2\omega$) are

$$-6 + 3\omega = -6 + 3\omega + (3 + \omega)(0),$$

$$-3 + 4\omega = -6 + 3\omega + (3 + \omega)(1),$$

$$-5 + 7\omega = -6 + 3\omega + (3 + \omega)\omega,$$

and

$$-2 + 8\omega = -6 + 3\omega + (3 + \omega)(1 + \omega).$$

Identical Polynomials Modulo μ

DEFINITION Two polynomials $f(X) = \alpha_0 + \alpha_1 X + \ldots + \alpha_n X^n$ and $g(X) = \beta_0 + \beta_1 X + \ldots + \beta_n X^n$ are said to be **identical modulo** μ if $\alpha_i \equiv \beta_i$ (MOD μ) for $i = 0, 1, 2, \ldots, n$. When $f(X)$ and $g(X)$ are identical modulo μ, then we write $f(X) \equiv g(X)$ (MOD μ).

THEOREM 6.8 *Let* $f(X)$, $g(X)$, *and* $h(X)$ *be polynomials over* $\hat{Z}$ *and* $F(X)$, $G(H)$, *and* $H(X)$ *the corresponding polynomials over* $\hat{Z}_\mu$.
(i) *If* $f(X) \equiv g(X)$ (MOD μ), *then* $f(\xi) \equiv g(\xi)$ (MOD μ) *for every integer* ξ.
(ii) $f(X) \equiv g(X)$ (MOD μ) *if and only if* $F(X) = G(X)$.
(iii) $f(X) \equiv g(X)h(X)$ (MOD μ) *if and only if* $F(X) = G(X)H(X)$.
PROOF: Statements (i) and (ii) are immediate consequences of the definition of identical modulo μ. Thus we concentrate our efforts on (iii). Let α_i, β_i, γ_i denote the coefficients of X^i in $f(X)$, $g(X)$, $h(X)$, respectively. Then the coefficient of X^i in $g(X)h(X)$ is

$$\beta_0\gamma_i + \beta_1\gamma_{i-1} + \ldots + \beta_i\gamma_0,$$

and the coefficient of X^i in $G(X)H(X)$ is

$$\beta_0{}^*\gamma_i{}^* + \beta_1{}^*\gamma_{i-1}{}^* + \ldots + \beta_i{}^*\gamma_0{}^*.$$

We will have established (iii) provided we can deduce that $\alpha_i \equiv \beta_0\gamma_i + \beta_1\gamma_{i-1} + \ldots + \beta_i\gamma_0$ (MOD μ) if and only if $\alpha_i{}^* = \beta_0{}^*\gamma_i{}^* + \beta_1{}^*\gamma_{i-1}{}^* + \ldots + \beta_i{}^*\gamma_0{}^*$. Since $\beta_0{}^*\gamma_i{}^* + \beta_1{}^*\gamma_{i-1}{}^* + \ldots + \beta_i{}^*\gamma_0{}^* = (\beta_0\gamma_i + \beta_1\gamma_{i-1} + \ldots + \beta_i\gamma_0)^*$, then the desired result follows from the fact that two integers are congruent modulo μ if and only if they are in the same residue class modulo μ. Q.E.D.

It should be noted that the converse of (i) in Theorem 6.8 is not true. For example, if π is a prime, then $f(X) = X^{|N(\pi)|}$ and $g(X) = X$ are not identical modulo π even though it follows from Fermat's theorem that $f(\xi) \equiv g(\xi)$ (MOD π) for every ξ.

Algebraic Congruences with Prime Moduli

If π is a prime, then $\hat{Z}_\pi$ is a field (Theorem 5.26), and so we may appeal to the standard theorems for polynomials with coefficients in a field when we are considering polynomials over $\hat{Z}_\pi$. Also, in light of Theorem 6.8, results for algebraic congruences modulo π can be deduced from the corresponding results for polynomials over the field $\hat{Z}_\pi$. There is no similar comprehensive theory when the modulus μ is not a prime. The specific results we shall need are rather meager and are contained in Theorems 6.9, 6.10, and 6.11.

THEOREM 6.9 *If π is a prime and $\alpha_n \not\equiv 0$ (MOD π), then the algebraic congruence*

$$f(X) = \alpha_0 + \alpha_1 X + \ldots + \alpha_n X^n \equiv 0 \ (\text{MOD } \pi)$$

has not more than n incongruent solutions modulo π.
PROOF: The polynomial $F(X) = \alpha_0{}^* + \alpha_1{}^*X + \ldots + \alpha_n{}^*X^n$ over the field $\hat{Z}_\pi$ has at most n roots in $\hat{Z}_\pi$. The result now follows from Corollary 6.4. Q.E.D.

DEFINITION The linear polynomial $X - \xi$ is said to be a **linear factor modulo** μ of the polynomial $f(X)$ over $\hat{Z}$ if $f(X) \equiv (X - \xi) \cdot g(X)$ (MOD μ) for some polynomial $g(X)$ over $\hat{Z}$.

THEOREM 6.10 (Remainder theorem) *If π is a prime, then $X - \xi$ is a linear factor modulo π of the polynomial $f(X)$ over $\hat{Z}$ if and only if $f(\xi) \equiv 0$ (MOD π).*

PROOF: If $X - \xi$ is a linear factor modulo π of $f(X)$, then $f(X) \equiv (X - \xi)g(X)$ (MOD π), and so $f(\xi) \equiv (\xi - \xi)g(\xi) = 0$ (MOD π), by Theorem 6.8(i).

Conversely, suppose that $f(\xi) \equiv 0$ (MOD π). Then, if $F(X)$ is the polynomial over $\hat{Z}_\pi$ corresponding to $f(X)$, we have $F(\xi^*) = 0^*$, by Corollary 6.2. Consequently, by the remainder theorem for polynomials over a field, we have $F(X) = (X - \xi^*)G(X)$, where $G(X) = \beta_0^* + \beta_1^* X + \ldots + \beta_r^* X^r$ is a polynomial over $\hat{Z}_\pi$. Hence, if $g(X) = \beta_0 + \beta_1 X + \ldots + \beta_r X^r$, then we obtain the result $f(X) \equiv (X - \xi)g(X)$ (MOD π), by Theorem 6.8(iii). Therefore $X - \xi$ is a linear factor modulo π of $f(X)$. Q.E.D.

THEOREM 6.11 *If π is a prime and the algebraic congruence $f(X) = \alpha_0 + \alpha_1 X + \ldots + \alpha_n X^n \equiv 0$ (MOD π) of degree n has r incongruent solutions, $\xi_1, \xi_2, \ldots, \xi_r$, then $f(X) \equiv (X - \xi_1)(X - \xi_2) \cdots (X - \xi_r)f_r(X)$ (MOD π), where $f_r(X)$ is a polynomial of degree $n - r$ over $\hat{Z}$ and highest coefficient α_n.*

PROOF: The residue classes $\xi_1^*, \xi_2^*, \ldots, \xi_r^*$ are distinct roots of $F(X) = \alpha_0^* + \alpha_1^* X + \ldots + \alpha_n^* X^n$, and so it follows from a standard result for polynomials over fields that $F(X) = (X - \xi_1^*)(X - \xi_2^*) \cdots (X - \xi_r^*)F_r(X)$, where $F_r(X) = \beta_0^* + \beta_1^* X + \ldots + \beta_{n-r-1}^* X^{n-r-1} + \alpha_n^* X^{n-r}$. Thus, by Theorem 6.8, we obtain $f(X) \equiv (X - \xi_1)(X - \xi_2) \cdots (X - \xi_r)f_r(X)$ (MOD π), where $f_r(X) = \beta_0 + \beta_1 X + \ldots + \beta_{n-r-1} X^{n-r-1} + \alpha_n X^{n-r}$. Q.E.D.

COROLLARY 6.12 *If π is a prime and the algebraic congruence $f(X) = \alpha_0 + \alpha_1 X + \ldots + \alpha_n X^n \equiv 0$ (MOD π) of degree n has n incongruent solutions, $\xi_1, \xi_2, \ldots, \xi_n$, then $f(X) \equiv \alpha_n(X - \xi_1)(X - \xi_2) \cdots (X - \xi_n)$ (MOD π).*

Theorems 6.9 and 6.11 are not true when the modulus is not a prime. For example, 1, 3, $1 + 2\omega$, and $3 + 2\omega$ are four incongruent solutions of the algebraic congruence $X^2 - 5 \equiv 0$ (MOD 4) of degree 2. In addition,

$$(X - 3)(X - (1 + 2\omega)) = X^2 - (4 + 2\omega)X + (3 + 6\omega),$$

and since $3 + 6\omega \equiv -5$ (MOD 4), then $X^2 - 5 \equiv (X - 3)(X - (1 + 2\omega))$ (MOD 4).

COROLLARY 6.13 *If π is a prime and $\{\xi_1, \xi_2, \ldots, \xi_r\}$, where $r = |N(\pi)| - 1$, is a reduced set of residues modulo π, then $X^r - 1 \equiv (X - \xi_1)(X - \xi_2) \cdots (X - \xi_r)$ (MOD π).*

PROOF: By Fermat's theorem, $\xi_i^r - 1 \equiv 0$ (MOD π) for $i = 1, 2, \ldots, r$. Consequently, $\xi_1, \xi_2, \ldots, \xi_r$ are r incongruent solutions of the algebraic congruence $X^r - 1 \equiv 0$ (MOD π). The result now follows from Corollary 6.12. Q.E.D.

THEOREM 6.14 (Wilson's theorem for $\hat{Z}$) *If π is a prime and $\{\xi_1, \xi_2, \ldots, \xi_r\}$, $r = |N(\pi)| - 1$, is a reduced set of residues modulo π, then $\xi_1 \xi_2 \cdots \xi_r \equiv -1$ (MOD π).*

PROOF: By Corollary 6.13, $X^r - 1 \equiv (X - \xi_1)(X - \xi_2) \cdots (X - \xi_r)$ (MOD π). Substituting 0 for X, we obtain that $(-1)^r \xi_1 \xi_2 \cdots \xi_r \equiv -1$ (MOD π). If π is an associate of 2, then $-1 \equiv 1$ (MOD π), whence $(-1)^r \equiv 1$ (MOD π), and thus $\xi_1 \xi_2 \cdots \xi_r \equiv -1$ (MOD π). If π is not an associate of 2, then $r = |N(\pi)| - 1$, and so $r = p - 1$ or $r = p^2 - 1$, where p is an odd rational prime. Thus r is even, $(-1)^r = 1$, and the result $\xi_1 \xi_2 \cdots \xi_r \equiv -1$ (MOD π) follows. Q.E.D.

Algebraic Congruences with Prime-Power Moduli

In this section we will be investigating the relationship between the algebraic congruences

$$f(X) \equiv 0 \ (\text{MOD} \ \pi^k), \tag{6.6}$$

$$f(X) \equiv 0 \ (\text{MOD} \ \pi^{k-1}), \tag{6.7}$$

where π is a prime and $k > 1$.

THEOREM 6.15 *If ξ is a solution of (6.6), then $f(\xi) = \alpha \pi^k$ for some α, and ξ is a solution of (6.7).*
PROOF: Trivial.

THEOREM 6.16 *If $\zeta_1, \zeta_2, \ldots, \zeta_r$ are the incongruent solutions of (6.7), then every solution ξ of (6.6) is of the form $\xi = \zeta_i + \eta \pi^{k-1}$ for some η and some i, $i = 1, 2, \ldots, r$.*
PROOF: If ξ is a solution of (6.6), it follows from Theorem 6.15 that ξ is a solution of (6.7). Consequently, $\xi \equiv \zeta_i \ (\text{MOD} \ \pi^{k-1})$ for some i. Thus there exists an η such that $\xi = \zeta_i + \eta \pi^{k-1}$. Q.E.D.

DEFINITION If ζ is a solution of (6.7) and $\xi = \zeta + \eta \pi^{k-1}$ is a solution of (6.6) for some η, then ξ is said to be a **solution of (6.6) corresponding to the solution ζ of (6.7).**

From Theorem 6.16 it is clear that if we can find the incongruent solutions $\zeta_1, \zeta_2, \ldots, \zeta_r$ of (6.7), and for each ζ_i find the solutions ξ of (6.6) corresponding to ζ_i, then we will have found all solutions of (6.6). Thus all that remains to be done is to establish an algorithm whereby the solutions ξ of (6.6) corresponding to a solution ζ of (6.7) can be determined. We begin this task with the easy

LEMMA 6.17 *If $f^{(k)}(X)$ is the kth derivative of $f(X) = \alpha_0 + \alpha_1 X + \ldots + \alpha_n X^n$ and ζ is an integer, then $f^{(k)}(\zeta)/k!$ is an integer.*
PROOF: If $k > n$, then $f^{(k)}(\zeta)/k! = 0$. Hence we may assume that $0 \leq k \leq n$. For these values of k we have

$$f^{(k)}(X) = \sum_{i=k}^{n} i(i-1) \cdots (i-k+1)\alpha_i X^{i-k}$$

$$= \sum_{i=k}^{n} k! \binom{i}{k} \alpha_i X^{i-k}.$$

Thus $f^{(k)}(\zeta)/k! = \sum_{i=k}^{n} \binom{i}{k} \alpha_i \zeta^{i-k}$, which is clearly an integer. Q.E.D.

THEOREM 6.18 *If* $f(\zeta) = \alpha \pi^{k-1}$, *then* $\xi = \zeta + \eta \pi^{k-1}$ *is a solution of the algebraic congruence* (6.6) *if and only if* $\alpha + \eta f'(\zeta) \equiv 0$ *(MOD* π).

PROOF: By Taylor's formula, we have

$$f(\xi) = f(\zeta + \eta \pi^{k-1})$$

$$= f(\zeta) + f'(\zeta)(\eta \pi^{k-1}) + \sum_{i=2}^{n} \frac{f^{(i)}(\zeta)}{i!} (\eta \pi^{k-1})^i,$$

where n is the degree of the polynomial $f(X)$. Since $f^{(i)}(\zeta)/i!$ is an integer for $i = 2, 3, \ldots, n$, by Lemma 6.17, then

$$\sum_{i=2}^{n} \frac{f^{(i)}(\zeta)}{i!} (\eta \pi^{k-1})^i = \theta \pi^{2k-2},$$

where θ is an integer. Furthermore, since $k \geq 2$, then $2k - 2 \geq k$, and so

$$\sum_{i=2}^{n} \frac{f^{(i)}(\zeta)}{i!} (\eta \pi^{k-1})^i \equiv 0 \ (\text{MOD } \pi^k).$$

Consequently, upon using the result $f(\zeta) = \alpha \pi^{k-1}$, we obtain $f(\xi)$ $\equiv \alpha \pi^{k-1} + f'(\zeta)(\eta \pi^{k-1}) = (\alpha + \eta f'(\zeta)) \pi^{k-1}$ (MOD π^k). Accordingly, $f(\xi) \equiv 0$ (MOD π^k) if and only if $(\alpha + \eta f'(\zeta)) \pi^{k-1} \equiv 0$ (MOD π^k). The last congruence is equivalent to $\alpha + \eta f'(\zeta) \equiv 0$ (MOD π), and the demonstration is complete. Q.E.D.

THEOREM 6.19 *If* ζ *is a solution of the algebraic congruence* (6.7), *then*
(i) *if* $f'(\zeta) \not\equiv 0$ (MOD π), *there is a unique solution modulo* π^k *of* (6.6) *corresponding to* ζ;
(ii) *if* $f'(\zeta) \equiv 0$ (MOD π), *there are* $|N(\pi)|$ *incongruent solutions modulo* π^k *of* (6.6) *corresponding to* ζ *when* $f(\zeta) \equiv 0$ (MOD π^k), *and no such solution when* $f(\zeta) \not\equiv 0$ (MOD π^k).

PROOF: Since ζ is a solution of (6.7), then $f(\zeta) = \alpha \pi^{k-1}$ for some α. If $f'(\zeta) \not\equiv 0$ (MOD π), then $\text{GCD}(f'(\zeta), \pi) = 1$, and therefore the linear congruence $\alpha + f'(\zeta)X \equiv 0$ (MOD π) has a unique solution η modulo π. Consequently, it follows from Theorem 6.18 that $\xi = \zeta + \eta \pi^{k-1}$ is the unique solution modulo π^k of (6.6) corresponding to ζ. If $f'(\zeta) \equiv 0$ (MOD π), then $\alpha + \eta f'(\zeta) \equiv \alpha$ (MOD π) for every integer η. Thus, if $\alpha \not\equiv 0$ (MOD π), then there is no solution of (6.6) corresponding to ζ, while if $\alpha \equiv 0$ (MOD π), then every $\xi = \zeta + \eta\pi^{k-1}$ is a solution. Since there are $|N(\pi)|$ integers in a complete set of residues modulo π, it is clear that we will have $|N(\pi)|$ solutions of (6.6) corresponding to ζ when $\alpha \equiv 0$ (MOD π). We conclude the proof by noting that since $f(\zeta) = \alpha\pi^{k-1}$, then $\alpha \equiv 0$ (MOD π) if and only if $f(\zeta) \equiv 0$ (MOD π^k). Q.E.D.

The following two examples illustrate how we may use the preceding results to solve algebraic congruences when the modulus is a power of a prime.

EXAMPLE 6.2   Let $f(X) = X^2 - 5$. A complete set of residues modulo 2 is $\{0, 1, \omega, \omega^2 = 1 + \omega\}$. By direct substitution, we find that 1 is the only solution modulo 2 of $f(X) \equiv 0$ (MOD 2). Noting that $f(1) = (-2)(2)$ and $f'(1) = (1)(2) \equiv 0$ (MOD 2), we deduce that there are $|N(2)| = 4$ incongruent solutions modulo 4 of $f(X) \equiv 0$ (MOD 4); and they are given by $\xi = 1 + 2\eta$, where η satisfies the congruence $-2 + 2\eta \equiv 0$ (MOD 2). Consequently, the four incongruent solutions of $X^2 - 5 \equiv 0$ (MOD 4) are $1 = 1 + (2)(0)$, $3 = 1 + (2)(1)$, $1 + 2\omega$, and $3 + 2\omega = 1 + 2(1 + \omega)$.

EXAMPLE 6.3   Let $f(X) = X^3 + X + \omega$. As in Example 6.2, we find, by direct substitution, that $\omega^2 = 1 + \omega$ is the only solution modulo 2 of $f(X) \equiv 0$ (MOD 2). Noting that $f(\omega^2) = (3 + 5\omega)(2)$ and $f'(\omega^2) = 3\omega^4 + 1 = 7 + 9\omega \not\equiv 0$ (MOD 2), we deduce that the unique solution modulo 4 of $f(X) \equiv 0$ (MOD 4) is given by $\xi = \omega^2 + 2\eta$, where η is the solution of the congruence $(3 + 5\omega) + (7 + 9\omega)X \equiv 0$ (MOD 2). By inspection, $\eta = 1$, and so $3 + \omega = \omega^2 + 2$ is the unique solution modulo 4 of the congruence $X^3 + X + \omega \equiv 0$ (MOD 4).

Algebraic Congruences with Composite Moduli

We bring the chapter to a close by showing how the Chinese remainder theorem can be coupled with previous results to solve the algebraic congruence $f(X) \equiv 0$ (MOD μ).

THEOREM 6.20 Let $\mu = \pi_1^{c_1}\pi_2^{c_2}\cdots\pi_r^{c_r}$, where the π_i are nonassociated primes and the c_i are positive. If $f(X)$ is a polynomial over $\hat{Z}$, and ξ_i is a solution of $f(X) \equiv 0$ (MOD $\pi_i^{c_i}$) for $i = 1, 2, \ldots, r$, then any integer ξ satisfying $\xi \equiv \xi_i$ (MOD $\pi_i^{c_i}$) for $i = 1, 2, \ldots, r$ is a solution of $f(X) \equiv 0$ (MOD μ). Furthermore, such a ξ exists, is unique modulo μ, and can be obtained by the Chinese remainder theorem.

PROOF: Since the $\pi_i^{c_i}$ are relatively prime in pairs, then, by the Chinese remainder theorem, there exists a unique ξ modulo μ such that $\xi \equiv \xi_i$ (MOD $\pi_i^{c_i}$) for $i = 1, 2, \ldots, r$. Thus, $f(\xi) \equiv f(\xi_i) \equiv 0$ (MOD $\pi_i^{c_i}$), $i = 1, 2, \ldots, r$, and since the $\pi_i^{c_i}$ are relatively prime in pairs, we have $f(\xi) \equiv 0$ (MOD μ). Accordingly, ξ is a solution of $f(X) \equiv 0$ (MOD μ). Q.E.D.

COROLLARY 6.21 Let $\mu = \pi_1^{c_1}\pi_2^{c_2}\cdots\pi_r^{c_r}$, where the π_i are nonassociated primes and the c_i are positive, and let $f(X)$ be a polynomial over $\hat{Z}$. Then
(i) the algebraic congruence

$$f(X) \equiv 0 \ (\text{MOD } \mu) \tag{6.8}$$

has a solution if and only if each of the algebraic congruences

$$f(X) \equiv 0 \ (\text{MOD } \pi_i^{c_i}), \tag{6.9}$$

$i = 1, 2, \ldots, r$, has a solution;
(ii) if $n(\mu)$ and $n(\pi_i^{c_i})$, $i = 1, 2, \ldots, r$, denote the number of incongruent solutions of (6.8) and (6.9), respectively, then

$$n(\mu) = n(\pi_1^{c_1}) \cdot n(\pi_2^{c_2}) \cdots n(\pi_r^{c_r}).$$

PROOF: If $f(\xi) \equiv 0$ (MOD μ), then $f(\xi) \equiv 0$ (MOD $\pi_i^{c_i}$) for $i = 1, 2, \ldots, r$. Hence, if ξ is a solution of (6.8), then ξ is a solution of (6.9). Conversely, if ξ_i is a solution of (6.9) for $i = 1, 2, \ldots, r$, then, by Theorem 6.20, there exists a unique solution ξ modulo μ of (6.8) such that $\xi \equiv \xi_i$ (MOD $\pi_i^{c_i}$) for $i = 1, 2, \ldots, r$. It follows that different solutions of (6.8) must arise from a different collection $\xi_1, \xi_2, \ldots, \xi_r$ of solutions of the congruences (6.9), and that different collections $\xi_1, \xi_2, \ldots, \xi_r$ will yield different solutions of (6.8). Hence all solutions of (6.8) will be obtained by allowing $\xi_1, \xi_2, \ldots, \xi_r$ to take all permissible values. Since for each i, ξ_i can take $n(\pi_i^{c_i})$ distinct values, the result (ii) follows. Q.E.D.

EXAMPLE 6.4 We will use Theorem 6.20 to find the solutions of the congruence $X^2 - 5 \equiv 0$ (MOD $12 + 4\omega$). By the division algorithm, we have $3 + \omega = 4(1) + (-1 + \omega) = 4 + \omega^{-1}$. Consequently, $\omega^{-1} = (3 + \omega) - 4$, so that $1 = (3 + \omega)\omega - 4\omega$. From this we conclude that $(3 + \omega)\omega \equiv 1$ (MOD 4) and $4(-\omega) \equiv 1$ (MOD $3 + \omega$). Thus the integer $\xi = -4\xi_1\omega + (3 + \omega)\omega\xi_2 = -4\omega\xi_1 + (1 + 4\omega)\xi_2$ is the unique simultaneous solution modulo $12 + 4\omega$ of the congruences $\xi \equiv \xi_1$ (MOD $3 + \omega$), $\xi \equiv \xi_2$ (MOD 4).

A complete set of residues modulo $3 + \omega$ is $\{0, 1, 2, \ldots, 10\}$. By direct substitution, we find that 4 and 7 are the incongruent solutions of $X^2 - 5 \equiv 0$ (MOD $3 + \omega$). Furthermore, in Example 6.2 we found that 1, 3, $1 + 2\omega$, and $3 + 2\omega$ were the incongruent solutions of $X^2 - 5 \equiv 0$ (MOD 4).

Using the information from the preceding two paragraphs, we construct the following table in which the last column gives the eight solutions of $X^2 - 5 \equiv 0$ (MOD $12 + 4\omega$).

ξ_1	ξ_2	$\xi = -4\omega\xi_1 + (1 + 4\omega)\xi_2$
4	1	$1 - 12\omega$
4	3	$3 - 4\omega$
4	$1 + 2\omega$	$9 - 2\omega$
4	$3 + 2\omega$	$11 + 6\omega$
7	1	$1 - 24\omega$
7	3	$3 - 16\omega$
7	$1 + 2\omega$	$9 - 14\omega$
7	$3 + 2\omega$	$11 - 6\omega$

In closing, we note that no generality is lost by taking μ to be of the form given in Theorem 6.20. For if μ_1 is an associate of μ, then the algebraic congruences $f(X) \equiv 0$ (MOD μ) and $f(X) \equiv 0$ (MOD μ_1) have precisely the same solutions; and any nonunit $\mu_1 \neq 0$ is an associate of some μ of the form given in Theorem 6.20. Indeed, Theorem 6.20 and Corollary 6.21 remain valid if μ is replaced by μ_1.

VII Primitive Roots

Introduction

Our primary objective in this chapter is to determine the integers μ such that, for some ρ, the set $\{1, \rho, \rho^2, \ldots, \rho^{\Phi(\mu)-1}\}$ is a reduced set of residues modulo μ. This is equivalent to determining the integers μ for which $\hat{U}_\mu$ is a cyclic group. Thus both the number-theorist and the algebraist will appreciate the importance of the task before us. The theoretical development is similar in many aspects to the corresponding development in rational integer theory. However, there are important differences and, as should be expected, some of the results are much harder to come by.

Order of α Modulo μ

DEFINITION If $GCD(\alpha, \mu) = 1$, then the **order of α modulo μ** is defined to be the smallest $r > 0$ such that $\alpha^r \equiv 1$ (MOD μ), and we write $r = \text{ord}(\alpha, \mu)$.

We observe that if $GCD(\alpha, \mu) = 1$, then $\alpha^{\Phi(\mu)} \equiv 1$ (MOD μ), by the Euler–Fermat theorem, so that $\text{ord}(\alpha, \mu)$ exists and $\text{ord}(\alpha, \mu) \leq \Phi(\mu)$. Also, if $\alpha^r \equiv 1$ (MOD μ) for some $r > 0$, then $1 = GCD(1, \mu) = GCD(\alpha^r, \mu) = GCD(\alpha, \mu)$. Thus, if $GCD(\alpha, \mu)$ is not a unit, then $\alpha^r \not\equiv 1$ (MOD μ) for every $r > 0$. Lastly, since $\alpha^r \equiv 1$ (MOD μ) if and only if $(\alpha^*)^r = 1^*$, we conclude that $\text{ord}(\alpha, \mu)$ is the order of the element α^* in the group of units $\hat{U}_\mu$. We record these results in

THEOREM 7.1 *If* $GCD(\alpha, \mu) = 1$, *then*
(i) $\text{ord}(\alpha, \mu) \leq \Phi(\mu)$;
(ii) $\text{ord}(\alpha, \mu)$ *is the order of the element* α^* *in the group of units* $\hat{U}_\mu$.

THEOREM 7.2 *If* $GCD(\alpha, \mu) = 1$ *and* $r = \text{ord}(\alpha, \mu)$, *then*
(i) $1, \alpha, \alpha^2, \ldots, \alpha^{r-1}$ *are incongruent modulo* μ;
(ii) *if* $n \geq 0$, *there exists a unique* s *in the range* $0 \leq s < r$ *such that* $\alpha^n \equiv \alpha^s$ (MOD μ);
(iii) *if* $n \geq 0$, *then* $\alpha^n \equiv 1$ (MOD μ) *if and only if* $r \mid n$;
(iv) *if* $n \geq 0$ *and* $m \geq 0$, *then* $\alpha^n \equiv \alpha^m$ (MOD μ) *if and only if* $n \equiv m$ (mod r);
(v) *if* $\beta \equiv \alpha$ (MOD μ), *then* $\text{ord}(\beta, \mu) = \text{ord}(\alpha, \mu)$;
(vi) *if* μ *and* μ_1 *are associates, then* $\text{ord}(\alpha, \mu) = \text{ord}(\alpha, \mu_1)$.
PROOF: The element α^* in $\hat{U}_\mu$ generates the cyclic subgroup $\{1^*, \alpha^*, (\alpha^2)^*, \ldots, (\alpha^{r-1})^*\}$ of order r. Since the elements $(\alpha^k)^*$, $k = 0, 1, 2, \ldots, r - 1$, are distinct, then (i) follows. If $n \geq 0$, then there exists a unique s in the range $0 \leq s < r$ such that $(\alpha^n)^* = (\alpha^s)^*$, or equivalently, $\alpha^n \equiv \alpha^s$ (MOD μ). Thus (ii) is established. From basic properties of finite groups, we have $(\alpha^n)^* = 1^*$ if and only if $r \mid n$. This establishes (iii), and (iv) is accomplished in a similar fashion, because $(\alpha^n)^* = (\alpha^m)^*$ if and only if $r \mid (n - m)$. Assertion (v) follows from Theorem 7.1(ii), since $\beta^* = \alpha^*$. Assertion (vi) follows from Theorem 7.1(ii) and the fact that $\hat{U}_{\mu_1} = \hat{U}_\mu$. Q.E.D.

COROLLARY 7.3 *If* $GCD(\alpha, \mu) = 1$ *and* $r = \text{ord}(\alpha, \mu)$, *then* $r \mid \Phi(\mu)$.
PROOF: We have $\Phi(\mu) > 0$ and $\alpha^{\Phi(\mu)} \equiv 1$ (MOD μ). Accordingly, $r \mid \Phi(\mu)$, by Theorem 7.2(iii). Q.E.D.

When finding ord(α, μ), Corollary 7.3 says that we need only check the positive rational integer divisors of $\Phi(\mu)$. We can frequently reduce the number of divisors of $\Phi(\mu)$ which must be checked. The following example shows how Theorem 5.51 may be used to good advantage in determining ord($2 + \omega, 13$).

EXAMPLE 7.1 By Theorem 5.51(i), we have $(2 + \omega)^{14} \equiv (2 + \omega) \cdot (2 + \bar{\omega}) = 5$ (MOD 13). Thus $(2 + \omega)^{56} \equiv 5^4 \equiv 1$ (MOD 13). Therefore ord($2 + \omega, 13$) divides $56 = 2^3 \cdot 7$, and so ord($2 + \omega, 13$) = 1, 2, 4, 7, 8, 14, 28, or 56. Now $(2 + \omega)^k$, $k = 1, 2, 4, 7, 8, 14, 28, 56$, is congruent modulo 13 to $2 + \omega, 5 + 5\omega, -2 - 3\omega, 1 - 2\omega, -5\omega, 5, -1, 1$, respectively. Consequently, ord($2 + \omega, 13$) = 56.

The following three theorems give some very useful tools for computing orders of elements.

THEOREM 7.4 *If $\mu_1, \mu_2, \ldots, \mu_s$ are nonzero pairwise relatively prime integers, GCD($\alpha, \mu_1 \mu_2 \cdots \mu_s$) = 1, and ord($\alpha, \mu_i$) = r_i for $i = 1, 2, \ldots, s$, then*

$$\text{ord}(\alpha, \mu_1 \mu_2 \cdots \mu_s) = \text{lcm}(r_1, r_2, \ldots, r_s).$$

PROOF: Let $t_1 = \text{ord}(\alpha, \mu_1 \mu_2 \cdots \mu_s)$ and $t_2 = \text{lcm}(r_1, r_2, \ldots, r_s)$. Then $\alpha^{t_1} \equiv 1$ (MOD $\mu_1 \mu_2 \cdots \mu_s$), and hence $\alpha^{t_1} \equiv 1$ (MOD μ_i) for $i = 1, 2, \ldots, s$. Thus $r_i \mid t_1$ for $i = 1, 2, \ldots, s$, by Theorem 7.2(iii). Accordingly, t_1 is a common multiple of the r_i, and so $t_2 \leq t_1$. Also, since $r_i \mid t_2$, then $\alpha^{t_2} \equiv 1$ (MOD μ_i) for $i = 1, 2, \ldots, s$, by Theorem 7.2(iii). But the μ_i are pairwise relatively prime, and so $\alpha^{t_2} \equiv 1$ (MOD $\mu_1 \mu_2 \cdots \mu_s$). Thus $t_1 \leq t_2$. Since we have shown that $t_1 \leq t_2$ and $t_2 \leq t_1$, then it follows that $t_1 = t_2$. Q.E.D.

THEOREM 7.5 *If ord(α, μ) = r_1, ord(β, μ) = r_2, and gcd(r_1, r_2) = 1, then ord($\alpha\beta, \mu$) = $r_1 r_2$.*
PROOF: Let $r = \text{ord}(\alpha\beta, \mu)$. Since $(\alpha\beta)^{r_1 r_2} = (\alpha^{r_1})^{r_2}(\beta^{r_2})^{r_1} \equiv 1$ (MOD μ), then $r \mid r_1 r_2$, by Theorem 7.2(iii). Thus $r \leq r_1 r_2$. Also, $1 \equiv (\alpha\beta)^{rr_1} = (\alpha^{r_1})^r \beta^{rr_1} \equiv \beta^{rr_1}$ (MOD μ), and $1 \equiv (\alpha\beta)^{rr_2} = \alpha^{rr_2}(\beta^{r_2})^r \equiv \alpha^{rr_2}$ (MOD μ). Consequently, $r_2 \mid rr_1$ and $r_1 \mid rr_2$. But gcd(r_1, r_2) = 1, and so $r_2 \mid r$, $r_1 \mid r$, and $r_1 r_2 \mid r$. Thus $r_1 r_2 \leq r$. Since we have shown that $r_1 r_2 \leq r$ and $r \leq r_1 r_2$, we conclude that $r = r_1 r_2$. Q.E.D.

THEOREM 7.6 *If* $\text{ord}(\alpha, \mu) = r$, $k \geq 0$, *and* $d = \gcd(r,k)$, *then* $\text{ord}(\alpha^k, \mu) = r/d$.

PROOF: Let $s = \text{ord}(\alpha^k, \mu)$. Since $(\alpha^k)^{r/d} = (\alpha^r)^{k/d} \equiv 1$ (MOD μ), then $s\,|(r/d)$, by Theorem 7.2(iii). But also $\alpha^{sk} = (\alpha^k)^s \equiv 1$ (MOD μ), and so $r\,|sk$, by the same theorem. Consequently, $(r/d)\,|s(k/d)$, and since $\gcd(r/d, k/d) = 1$, then $(r/d)\,|s$. From that and $s\,|(r/d)$ we conclude that $s = r/d$. Q.E.D.

Primitive Roots

DEFINITIONS If $r > 0$, then an integer ρ is said to be an **rth root** of an integer α modulo μ if $\rho^r \equiv \alpha$ (MOD μ). The integer ρ is said to be a **root of 1 modulo** μ if there exists an $r > 0$ such that ρ is an rth root of 1 modulo μ. The integer ρ is said to be a **primitive root modulo** μ (or a **primitive root of** μ) if every root of 1 modulo μ is congruent modulo μ to a positive rational integer power of ρ; that is, if α is a root of 1 modulo μ, then $\rho^r \equiv \alpha$ (MOD μ) for some $r > 0$. We say that the modulus μ **has a primitive root** if there exists a ρ such that ρ is a primitive root modulo μ.

The definition of primitive root was chosen to motivate the terminology and is not the usual number-theoretic definition. The customary definition is given by the following theorem.

THEOREM 7.7 *The integer ρ is a primitive root modulo μ if and only if* $\text{ord}(\rho, \mu) = \Phi(\mu)$.

PROOF: Suppose first that ρ is a primitive root modulo μ. Now 1 is a root of 1 modulo μ, and so $\rho^s \equiv 1$ (MOD μ) for some $s > 0$. Consequently, $\text{ord}(\rho, \mu)$ exists, say $r = \text{ord}(\rho, \mu)$. The element ρ^* of $\hat{U}_\mu$ generates the cyclic subgroup $\{1^*, \rho^*, (\rho^2)^*, \ldots, (\rho^{r-1})^*\}$ of order r. If $r \neq \Phi(\mu)$, then there exists an α^* in $\hat{U}_\mu$ such that α^* is not an element of the subgroup generated by ρ^*. Consequently, $(\rho^m)^* \neq \alpha^*$ for every $m > 0$, or equivalently, $\rho^m \not\equiv \alpha$ (MOD μ) for every $m > 0$. However α is a root of 1 modulo μ, since $\alpha^{\Phi(\mu)} \equiv 1$ (MOD μ). Thus, if $r \neq \Phi(\mu)$, then we would have a root α of 1 modulo μ that is not congruent modulo μ to a positive rational integer power of ρ. This contradiction forces us to conclude that $\text{ord}(\rho, \mu) = r = \Phi(\mu)$ whenever ρ is a primitive root modulo μ.

Conversely, suppose $\text{ord}(\rho, \mu) = \Phi(\mu)$. Then the $\Phi(\mu)$ integers 1, $\rho, \rho^2, \ldots, \rho^{\Phi(\mu)-1}$ are incongruent modulo μ, by Theorem 7.2(i), and thus constitute a reduced set of residues modulo μ. Therefore, if α is a root of 1, then $\text{GCD}(\alpha, \mu) = \text{GCD}(1, \mu) = 1$, and so there exists an s in the range $0 < s \leq \Phi(\mu)$ such that $\alpha \equiv \rho^s$ (MOD μ). Accordingly, if $\text{ord}(\rho, \mu) = \Phi(\mu)$, then ρ is a primitive root modulo μ. Q.E.D.

COROLLARY 7.8 *The integer ρ is a primitive root modulo μ if and only if the group $\hat{U}_\mu$ is cyclic with generator ρ^*.*
PROOF: The integer ρ is a primitive root modulo μ if and only if $\text{ord}(\rho, \mu) = \Phi(\mu)$. Also, $\text{ord}(\rho, \mu) = \Phi(\mu)$ if and only if the element ρ^* in $\hat{U}_\mu$ has order $\Phi(\mu)$. But ρ^* in $\hat{U}_\mu$ has order $\Phi(\mu)$ if and only if ρ^* is a generator of $\hat{U}_\mu$. Q.E.D.

COROLLARY 7.9 *The integer ρ is a primitive root modulo μ if and only if $\{1, \rho, \rho^2, \ldots, \rho^{\Phi(\mu)-1}\}$ is a reduced set of residues modulo μ.*
PROOF: The proof is immediate from Corollary 7.8, since $\{1, \rho, \rho^2, \ldots, \rho^{\Phi(\mu)-1}\}$ is a reduced set of residues if and only if $\{1^*, \rho^*, (\rho^2)^*, \ldots, (\rho^{\Phi(\mu)-1})^*\} = \hat{U}_\mu$. Q.E.D.

We interrupt the theoretical development for the following simple example in which we use Theorems 7.5 and 7.6 to show that $6 + 3\omega$ is a primitive root modulo 13.

EXAMPLE 7.2 From rational integer tables or by direct calculation, we see that $\text{ord}(2, 13) = \phi(13) = 12$. Now $\gcd(12, 4) = 4$, and so $\text{ord}(2^4, 13) = 12/4 = 3$, by Theorem 7.6. Since $2^4 \equiv 3$ (MOD 13), then we conclude that $\text{ord}(3, 13) = 3$. Recall from Example 7.1 that $\text{ord}(2 + \omega, 13) = 56$. Thus, since $\gcd(3, 56) = 1$, we have $\text{ord}(3(2 + \omega), 13) = (56)(3) = 168 = \Phi(13)$, by Theorem 7.5. Accordingly, $6 + 3\omega$ is a primitive root modulo 13.

Integers with Primitive Roots

When an integer μ has a primitive root ρ, then, by Corollary 7.9, the set $\{1, \rho, \rho^2, \ldots, \rho^{\Phi(\mu)-1}\}$ is a reduced set of residues modulo μ; that is, every integer relatively prime to μ is congruent modulo μ to a power of ρ. As we shall see in later investigations, this property is an exceedingly important theoretical tool. Thus it is advantageous to determine which integers μ have primitive roots. This is not an easy assignment, and much of this chapter is devoted toward classifying the integers μ having primitive roots. Ultimately the classification will be obtained in Theorem 7.23.

The first step in the classification is to show that every prime π has a primitive root. This result will follow as a simple corollary of Theorem 7.10, a theorem which is of obvious importance in its own right. The proof of this theorem is very similar to the corresponding analogue in rational integer number theory. In the proof we shall need the rational integer analogue of Theorem 5.49; that is, if $n > 0$, then $\sum_{d \mid n} \phi(d) = n$, where the sum is extended over all the positive rational integer divisors of n.

THEOREM 7.10 *If π is a prime and $d > 0$ is a divisor of $\Phi(\pi)$, then the algebraic congruence*

$$X^d - 1 \equiv 0 \ (\text{MOD } \pi)$$

has exactly d incongruent solutions, and exactly $\phi(d)$ of these solutions have order d modulo π.

PROOF: Since $d \mid \Phi(\pi)$, we have $X^{\Phi(\pi)} - 1 = (X^d - 1)g(X)$, where $g(X) = 1 + X^d + X^{2d} + \ldots + X^{\Phi(\pi)-d}$. By Corollary 6.13, the algebraic congruence $X^{\Phi(\pi)} - 1 \equiv 0 \ (\text{MOD } \pi)$ has $\Phi(\pi)$ incongruent solutions. If ξ is one of them, then $(\xi^d - 1)g(\xi) \equiv 0 \ (\text{MOD } \pi)$. Therefore either $\xi^d - 1 \equiv 0 \ (\text{MOD } \pi)$ or $g(\xi) \equiv 0 \ (\text{MOD } \pi)$. Thus each of the $\Phi(\pi)$ incongruent solutions of $X^{\Phi(\pi)} - 1 \equiv 0 \ (\text{MOD } \pi)$ is either a solution of $X^d - 1 \equiv 0 \ (\text{MOD } \pi)$ or of $g(X) \equiv 0 \ (\text{MOD } \pi)$. But, by Theorem 6.9, these congruences have, respectively, not more than d and $\Phi(\pi) - d$ incongruent solutions. Thus, since $\Phi(\pi) = d + \Phi(\pi) - d$, it follows that $X^d - 1 \equiv 0 \ (\text{MOD } \pi)$ has exactly d incongruent solutions.

Suppose now that one of these solutions, ξ say, has order d modulo π. Then $1, \xi, \xi^2, \ldots, \xi^{d-1}$ are all solutions of $X^d - 1 \equiv 0$ (MOD π) and, by Theorem 7.2(i), are incongruent modulo π. Thus they form a complete set of solutions modulo π of $X^d - 1 \equiv 0$ (MOD π). It follows from Theorem 7.6 that $\text{ord}(\xi^k, \pi) = d$ if and only if $\gcd(d, k) = 1$. Furthermore, there are precisely $\phi(d)$ values of k in the range $0 \leq k \leq d - 1$ such that $\gcd(k, d) = 1$. Consequently, we deduce that if the algebraic congruence $X^d - 1 \equiv 0$ (MOD π) has a solution ξ of order d modulo π, then it has exactly $\phi(d)$ incongruent solutions of order d modulo π. The result just proved can be stated as follows: If $\chi(d)$ denotes the number of incongruent solutions of $X^d - 1 \equiv 0$ (MOD π) of order d modulo π, then either $\chi(d) = 0$ or $\chi(d) = \phi(d)$. We now show that, in fact, $\chi(d) = \phi(d)$.

Since there are exactly $\Phi(\pi)$ incongruent integers prime to π, and since each of these has as order modulo π a positive rational integer divisor of $\Phi(\pi)$, we have

$$\sum_{d \,|\, \Phi(\pi)} \chi(d) = \Phi(\pi),$$

the sum extending over all positive rational integer divisors d of $\Phi(\pi)$. But, by the rational integer analogue of Theorem 5.49,

$$\sum_{d \,|\, \Phi(\pi)} \phi(d) = \Phi(\pi).$$

Therefore, since $\chi(d) \leq \phi(d)$ for each d, we conclude from the preceding two equations that $\chi(d) = \phi(d)$ for each d, and the theorem is proved. Q.E.D.

COROLLARY 7.11 *If π is a prime, then π has $\phi(\Phi(\pi))$ incongruent primitive roots.*
PROOF: Let $d = \Phi(\pi)$ in Theorem 7.10. Then we deduce that there are exactly $\phi(\Phi(\pi))$ incongruent integers modulo π having order $\Phi(\pi)$ modulo π; that is, π has $\phi(\Phi(\pi))$ incongruent primitive roots. Q.E.D.

Corollary 7.11 guarantees that every prime π has a primitive root, but the theorem which led to this corollary was an existence theorem and gave no clue as to how a primitive root could be ascertained. We now turn our attention to the task of determining a primitive root for a prime π. The analysis separates into two cases: (i) π is a prime such that $|N(\pi)| = q$, where q is a rational prime, and (ii) π is an associate of a rational prime p of the form $p \equiv \pm 2 \pmod 5$. Before proceeding, a few observations should be made. First, if q is a rational prime and $\gcd(g,q) = 1$, then, from rational integer number theory, we know that $\text{ord}(g,q)$ divides $\phi(q) = q - 1$. Secondly, from readily available rational integer tables, we can exhibit for small rational primes q a positive g such that $\text{ord}(g,q) = q - 1$. The second of these observations and the following theorem effectively handle the case when $|N(\pi)| = q$.

THEOREM 7.12 *If π is a prime such that $|N(\pi)| = q$, where q is a rational prime, $\gcd(g,q) = 1$, and $r \geq 0$, then $g^r \equiv 1 \pmod{\pi}$ if and only if $g^r \equiv 1 \pmod q$. Consequently, $\text{ord}(g,\pi) = \text{ord}(g,q)$, and g is a primitive root of π whenever $\text{ord}(g,q) = q - 1$.*
PROOF: Clear from Theorem 5.16. Q.E.D.

EXAMPLE 7.3 From tables, we find that $\text{ord}(2,5) = 4$ and $\text{ord}(6, 41) = 40$. Since $N(2 + \omega) = 5$ and $N(6 + \omega) = N(6 + \bar{\omega}) = 41$, then 2 is a primitive root of $2 + \omega$, and 6 is a primitive root of both $6 + \omega$ and $6 + \bar{\omega}$.

There remains the more troublesome case when π is an associate of a rational prime p of the form $p \equiv \pm 2 \pmod 5$. The following theorem substantially reduces the number of integers in a reduced set of residues which need to be checked in order to locate a primitive root.

THEOREM 7.13 *Let π be an associate of a rational prime p of the form $p \equiv \pm 2 \pmod 5$, and α an integer which is relatively prime to π. If $\text{ord}(N(\alpha), p) = r$, then $\text{ord}(\alpha,\pi) | r(p + 1)$. In particular, if $r < p - 1$, then α is not a primitive root of π.*
PROOF: By Theorem 5.51(i), $\alpha^p \equiv \bar{\alpha} \pmod p$, and so $\alpha^{p+1} \equiv \alpha\bar{\alpha} = N(\alpha) \pmod p$. Thus $1 \equiv (N(\alpha))^r \equiv \alpha^{r(p+1)} \pmod p$. Accordingly, $\text{ord}(\alpha,\pi) = \text{ord}(\alpha, p)$ divides $r(p + 1)$. For the in particular part, we note that if $r < p - 1$, then $\text{ord}(\alpha,\pi) < (p - 1)(p + 1) = p^2 - 1 = \Phi(\pi)$, and so α is not a primitive root of π. Q.E.D.

It follows from Theorem 7.13 that if π is an associate of a rational prime p of the form $p \equiv \pm 2$ (mod 5), then we may restrict our search for primitive roots of π to the integers α in a reduced set of residues for which $\operatorname{ord}(N(\alpha), p) = p - 1$. Once we have found and α such that $\operatorname{ord}(N(\alpha), p) = p - 1$, then there remains the task of determining whether the α is a primitive root of π. The following theorem reduces the number of congruences which must be checked in order to make this determination.

THEOREM 7.14 *Let π be an associate of a rational prime p of the form $p \equiv \pm 2$ (mod 5) and α an integer such that $\operatorname{ord}(N(\alpha), p) = p - 1$. Then $\operatorname{ord}(\alpha, \pi) = (p - 1)s$, where $s \mid (p + 1)$.*

PROOF: Let $r = \operatorname{ord}(\alpha, \pi)$. It suffices to show that $r \mid (p^2 - 1)$ and $(p - 1) \mid r$. It is immediate, by Corollary 7.3, that r divides $\Phi(\pi) = p^2 - 1$. Since $\alpha^r \equiv 1$ (MOD p), it is clear that $\bar{\alpha}^r \equiv 1$ (MOD p). Thus $(N(\alpha))^r = (\alpha\bar{\alpha})^r \equiv 1$ (MOD p). Consequently, since $\operatorname{ord}(N(\alpha), p) = p - 1$, then $(p - 1) \mid r$. Q.E.D.

EXAMPLE 7.4 We find a primitive root of the prime 13 using the results found in the preceding two theorems. Now $N(3 + \omega) = 11 \equiv -2$ (mod 13), and it is easily verified that $\operatorname{ord}(-2, 13) = 12$, so that $\operatorname{ord}(N(3 + \omega), 13) = 12$. Consequently, if $r = \operatorname{ord}(3 + \omega, 13)$, then it follows from Theorem 7.14 that $r = 12s$, where $s \mid 14$. From this we conclude that $\operatorname{ord}(3 + \omega, 13) = r = 12, 24, 84$, or 168. The following congruences modulo 13 are easily checked: $(3 + \omega)^{12} \equiv -2 - 3\omega$, $(3 + \omega)^{24} \equiv (-2 - 3\omega)^2 \equiv 8\omega$, $(3 + \omega)^{84} \equiv (8\omega)^3(-2 - 3\omega) \equiv -1$, and $(3 + \omega)^{168} \equiv (-1)^2 \equiv 1$. Therefore $\operatorname{ord}(3 + \omega, 13) = 168 = \Phi(13)$, which establishes that $3 + \omega$ is a primitive root of 13.

The method used in Example 7.4 is not without its drawbacks. For example, it is not difficult to show that $\operatorname{ord}(N(6 + \omega), 13) = 12$ and $\operatorname{ord}(6 + \omega, 13) = 24$. Consequently, $6 + \omega$ is not a primitive root modulo 13. There is an algorithm in rational integer theory due to Gauss for computing a g such that $\operatorname{ord}(g, p) = \phi(p)$. By suitably modifying this algorithm, it is possible to develop an algorithm for computing a primitive root modulo π, where π is a prime. However the computations required are quite formidable and are not suitable for hand computation. A short table of primitive roots modulo π is found in Appendix C. The table was constructed in the manner of Examples 7.3 and 7.4.

We now continue the task of classifying the integers μ with primitive roots. A sequence of eight lemmas will be established in order to obtain the classification in Theorem 7.23. Lemmas 7.17, 7.20, 7.21, and 7.22 may be considered technical lemmas whose purpose is to lay the groundwork for Theorem 7.23. Lemmas 7.15, 7.16, 7.18, and 7.19, on the other hand, have an importance apart from the immediate goal of proving Theorem 7.23 and should be considered an integral part of the theory of this chapter.

LEMMA 7.15 *If μ_1 and μ_2 are nonzero relatively prime integers, $GCD(\rho, \mu_1\mu_2) = 1$, $r_1 = \text{ord}(\rho, \mu_1)$, and $r_2 = \text{ord}(\rho, \mu_2)$, then ρ is a primitive root modulo $\mu_1\mu_2$ if and only if*
(i) *ρ is a primitive root modulo μ_1;*
(ii) *ρ is a primitive root modulo μ_2;*
(iii) *$\gcd(r_1, r_2) = 1$.*

PROOF: Suppose that ρ is a primitive root modulo $\mu_1\mu_2$. Then $\Phi(\mu_1) \cdot \Phi(\mu_2) = \Phi(\mu_1\mu_2) = \text{ord}(\rho, \mu_1\mu_2)$. Also, by Corollary 7.3, $r_1 | \Phi(\mu_1)$ and $r_2 | \Phi(\mu_2)$. Using these results and Theorem 7.4, we obtain $\Phi(\mu_1) \cdot \Phi(\mu_2) = \text{lcm}(r_1, r_2) \leq r_1 r_2 \leq \Phi(\mu_1) \cdot \Phi(\mu_2)$, so that $\Phi(\mu_1) \cdot \Phi(\mu_2) = \text{lcm}(r_1, r_2) = r_1 r_2$. Thus we have $r_1 | \Phi(\mu_1)$, $r_2 | \Phi(\mu_2)$, and $r_1 r_2 = \Phi(\mu_1) \cdot \Phi(\mu_2)$. Therefore $r_1 = \Phi(\mu_1)$, $r_2 = \Phi(\mu_2)$, and since $r_1 r_2 = \text{lcm}(r_1, r_2) = r_1 r_2/\gcd(r_1, r_2)$, then $\gcd(r_1, r_2) = 1$. Thus (i), (ii), and (iii) follow whenever ρ is a primitive root modulo $\mu_1\mu_2$. Conversely, suppose that (i), (ii), and (iii) are satisfied. Then $r_1 = \Phi(\mu_1)$, $r_2 = \Phi(\mu_2)$, and $\text{ord}(\rho, \mu_1\mu_2) = \text{lcm}(r_1, r_2) = r_1 r_2 = \Phi(\mu_1) \cdot \Phi(\mu_2) = \Phi(\mu_1\mu_2)$. Consequently, ρ is a primitive root modulo $\mu_1\mu_2$. Q.E.D.

LEMMA 7.16 *If the nonzero integer μ is neither a unit nor an associate of 2, then $2 | \Phi(\mu)$.*

PROOF: Either $\mu = \epsilon \cdot 2^k$ or $\mu = \mu_1\pi^j$, where ϵ is a unit, $j \geq 1$, $k \geq 2$, $GCD(\mu_1, \pi) = 1$, and π is a prime which is not an associate of 2. We shall show that $2 | \Phi(\mu)$ in each case. If $\mu = \epsilon \cdot 2^k$, $k \geq 2$, then $\Phi(\mu) = |N(2)|^{k-1} \cdot (|N(2)| - 1) = 4^{k-1} \cdot 3$, and thus $2 | \Phi(\mu)$. Now suppose that $\mu = \mu_1\pi^j$. Since $|N(\pi)| - 1 = p - 1$ or $p^2 - 1$, where p is an odd rational prime, then 2 divides $|N(\pi)| - 1$. Hence, $2 | \Phi(\mu)$, since $\Phi(\mu) = \Phi(\mu_1) \cdot \Phi(\pi^j) = \Phi(\mu_1) \cdot |N(\pi)|^{j-1} \cdot (|N(\pi)| - 1)$. Q.E.D.

LEMMA 7.17 *If the nonzero integer μ has a primitive root, then μ is a unit, or else there exist $k \geq 1$, a unit ϵ, and a prime π such that $\mu = \epsilon\pi^k$ or $\mu = (2\epsilon)\pi^k$. In the event that $\mu = (2\epsilon)\pi^k$, then we may assume that π is not an associate of 2.*

PROOF: Suppose to the contrary, that the nonzero integer μ is not one of the types stated. Then μ has at least two nonassociated prime factors, one of which, say π, is not an associate of 2. Thus $\mu = \mu_1\pi^k$, where μ_1 is not a unit nor an associate of 2, $k > 0$, and $\text{GCD}(\mu_1, \pi^k) = 1$. By the preceding lemma, we have $2 \mid \Phi(\mu_1)$ and $2 \mid \Phi(\pi^k)$. But if ρ is a primitive root modulo μ, it follows from Lemma 7.15 that we must have $\gcd(\Phi(\mu_1), \Phi(\pi^k)) = 1$, which is not true, as 2 divides both $\Phi(\mu_1)$ and $\Phi(\pi^k)$. This contradiction shows the lemma to be true. Q.E.D.

LEMMA 7.18 *If π is an associate of $2 + \omega$, $k \geq 3$, and $\text{GCD}(\alpha, \pi^k) = 1$, then*

$$\alpha^{\Phi(\pi^k)/5} \equiv 1 \ (\text{MOD } \pi^k).$$

PROOF: We induce on k. When $k = 3$ we have $\Phi(\pi^3) = |N(\pi^2)| \cdot (|N(\pi)| - 1) = 25(4) = 100$, so that $\alpha^{\Phi(\pi^3)/5} = \alpha^{20}$. Since $\{1, 2, 3, 4\}$ is a reduced set of residues modulo π, then $\alpha = j + \pi\beta$ for some j in $\{1, 2, 3, 4\}$ and some β. Thus $\alpha^{20} = (j + \pi\beta)^{20} \equiv j^{20} + 20j^{19}(\pi\beta) + 190j^{18}(\pi\beta)^2 \ (\text{MOD } \pi^3)$. Since $\sqrt{5}$ is an associate of π, then

$$20j^{19}(\pi\beta) = 4j^{19}\left(\sqrt{5}\right)^2\pi\beta \equiv 0 \ (\text{MOD } \pi^3)$$

and

$$190j^{18}(\pi\beta)^2 = 38j^{18}\left(\sqrt{5}\right)^2\pi^2\beta^2 \equiv 0 \ (MOD \ \pi^3).$$

Accordingly, $\alpha^{20} \equiv j^{20} \ (\text{MOD } \pi^3)$. But $\phi(5^2) = 5(4) = 20$, and so, by rational integer theory, $j^{20} \equiv 1 \ (\text{mod } 25)$. Hence, since $\pi^3 \mid 25$, then $\alpha^{20} \equiv j^{20} \equiv 1 \ (\text{MOD } \pi^3)$. Consequently, the formula is true for $k = 3$.

If now $\alpha^{\Phi(\pi^k)/5} \equiv 1 \ (\text{MOD } \pi^k)$, then $\alpha^{\Phi(\pi^k)/5} = 1 + \beta\pi^k$ for some β. Therefore

$$\alpha^{\Phi(\pi^{k+1})/5} = \alpha^{5^{k-1}(4)} = \alpha^{\Phi(\pi^k)}$$
$$= \left(1 + \beta\pi^k\right)^5 \equiv 1 + 5\beta\pi^k \ (\text{MOD } \pi^{k+1}).$$

But $5\beta\pi^k = (\sqrt{5})^2\pi^k\beta \equiv 0$ (MOD π^{k+1}). Consequently, $\alpha^{\Phi(\pi^{k+1})/5} \equiv 1$ (MOD π^{k+1}), and the induction is complete. Q.E.D.

LEMMA 7.19 *If π is an associate of a rational prime p of the form $p \equiv \pm 2$ (mod 5), $k \geq 2$, and $GCD(\alpha, \pi^k) = 1$, then $\alpha^{\Phi(\pi^k)/p} \equiv 1$ (MOD π^k).*

PROOF: We induce on k. When $k = 2$ we have $\Phi(\pi^2) = p^2(p^2 - 1)$, so that $\alpha^{\Phi(\pi^2)/p} = \alpha^{p(p^2-1)}$. Now

$$\left(\alpha^{p^2-1} - 1\right)^p = \alpha^{p(p^2-1)} + (-1)^p + \sum_{n=1}^{p-1} (-1)^{p-n}\binom{p}{n}\alpha^{n(p^2-1)}$$

and

$$0 = (1-1)^p = 1 + (-1)^p + \sum_{n=1}^{p-1} (-1)^{p-n}\binom{p}{n}.$$

Subtracting the two displayed equations, we obtain

$$\left(\alpha^{p^2-1} - 1\right)^p = \left(\alpha^{p(p^2-1)} - 1\right) + T, \tag{7.1}$$

where

$$T = \sum_{n=1}^{p-1} (-1)^{p-n}\binom{p}{n}\left[\alpha^{n(p^2-1)} - 1\right].$$

By the Euler–Fermat theorem, $\alpha^{p^2-1} = \alpha^{\Phi(p)} \equiv 1$ (MOD p). Consequently, $\alpha^{n(p^2-1)} \equiv 1$ (MOD p) for $n = 1, 2, \ldots, p - 1$. Since, in addition, $p \mid \binom{p}{n}$ for $n = 1, 2, \ldots, p - 1$, then we readily deduce that p^2 divides both T and $(\alpha^{p^2-1} - 1)^p$. Therefore we conclude from (7.1) that $\alpha^{p(p^2-1)} - 1 \equiv 0$ (MOD p^2), and hence $\alpha^{\Phi(\pi^2)/p} = \alpha^{p(p^2-1)} \equiv 1$ (MOD π^2). Thus the formula is true for $k = 2$.

If now $\alpha^{\Phi(\pi^k)/p} \equiv 1$ (MOD π^k), then $\alpha^{\Phi(\pi^k)/p} = 1 + p^k\beta$ for some β. Hence

$$\alpha^{\Phi(\pi^{k+1})/p} = \alpha^{p^{2k-1}(p^2-1)} = \alpha^{p\Phi(\pi^k)}$$

$$= \left(1 + p^k\beta\right)^{p^2} \equiv 1 \text{ (MOD } \pi^{k+1}),$$

completing the induction and the proof. Q.E.D.

LEMMA 7.20 *Let π be a prime such that $|N(\pi)| = q$, where q is a rational prime of the form $q = \pm 1$ (mod 5). Then there exists a primitive root ρ modulo π such that*

$$\rho^{q^{k-2}(q-1)} \not\equiv 1 \ (\mathrm{MOD}\ \pi^k)$$

for all $k \geq 2$.

PROOF: Let ρ be a primitive root modulo π. Then $\rho + \pi$ is also a primitive root modulo π, and since $\pi \mid q$, we obtain

$$(\rho + \pi)^{q-1} - \rho^{q-1} \equiv (q-1)\rho^{q-2}\pi \equiv -\rho^{q-2}\pi \ (\mathrm{MOD}\ \pi^2).$$

Accordingly, since $\mathrm{GCD}(\rho,\pi) = 1$, then

$$\left\{(\rho + \pi)^{q-1} - 1\right\} - \left\{\rho^{q-1} - 1\right\} \not\equiv 0 \ (\mathrm{MOD}\ \pi^2).$$

Thus not both ρ and $\rho + \pi$ can be solutions of the algebraic congruence $X^{q-1} \equiv 1 \ (\mathrm{MOD}\ \pi^2)$. We may therefore suppose that ρ has been chosen so that $\rho^{q-1} \not\equiv 1 \ (\mathrm{MOD}\ \pi^2)$. This establishes the result for $k = 2$.

Now suppose that $\rho^{q^{k-2}(q-1)} \not\equiv 1 \ (\mathrm{MOD}\ \pi^k)$. By the Euler–Fermat theorem, we have $\rho^{q^{k-2}(q-1)} = \rho^{\Phi(\pi^{k-1})} \equiv 1 \ (\mathrm{MOD}\ \pi^{k-1})$. Thus, by the preceding two results, we have $\rho^{q^{k-2}(q-1)} = 1 + \beta\pi^{k-1}$, where $\mathrm{GCD}(\beta,\pi) = 1$. Hence $\rho^{q^{k-1}(q-1)} = (1 + \beta\pi^{k-1})^q$ $\equiv 1 + q\beta\pi^{k-1} + \frac{1}{2}q(q-1)\beta^2\pi^{2k-2} \ (\mathrm{MOD}\ \pi^{k+1})$. Furthermore, since $\pi \mid q$, then π^{k+1} divides $\frac{1}{2}q(q-1)\beta^2\pi^{2k-2}$, and so $\rho^{q^{k-1}(q-1)}$ $\equiv 1 \pm \beta\bar{\pi}\pi^k \ (\mathrm{MOD}\ \pi^{k+1})$. But π and $\bar{\pi}$ are nonassociated primes, and so $\mathrm{GCD}(\beta\bar{\pi},\pi) = 1$. Therefore $\rho^{q^{k-1}(q-1)} \not\equiv 1 \ (\mathrm{MOD}\ \pi^{k+1})$, completing the induction and the proof. Q.E.D.

LEMMA 7.21 *Let π be a prime such that $|N(\pi)| = q$, where q is a rational prime of the form $q \equiv \pm 1 \ (\mathrm{mod}\ 5)$. If ϵ is a unit and $k \geq 1$, then $\mu = \epsilon\pi^k$ has a primitive root. If, in addition, $q \equiv -1 \ (mod\ 3)$, then $\mu_1 = (2\epsilon)\pi^k$ has a primitive root.*

PROOF: It suffices to show that π^k and $2\pi^k$ have primitive roots under the specified conditions. According to Lemma 7.20, there exists a primitive root ρ modulo π such that $\rho^{q^{k-2}(q-1)} \not\equiv 1 \ (\mathrm{MOD}\ \pi^k)$ for all $k \geq 2$. Let $r = \mathrm{ord}(\rho, \pi^k)$. We aim to show that $r = \Phi(\pi^k)$ for all $k \geq 1$, thus showing that ρ is a primitive root modulo π^k. When $k = 1$ the result is at hand, since ρ is a primitive root modulo π. Hence we may assume that $k \geq 2$. We know that r divides $\Phi(\pi^k) = q^{k-1}(q-1)$, by Corollary 7.3. Also, since $\rho^r \equiv 1 \ (\mathrm{MOD}\ \pi^k)$, then $\rho^r \equiv 1 \ (\mathrm{MOD}\ \pi)$, so that $q - 1 = \Phi(\pi) = \mathrm{ord}(\rho, \pi)$ divides r, by Theorem 7.2(iii). Consequently, we have $r = q^s(q-1)$, where $0 \leq s \leq k - 1$. If $r \neq q^{k-1}(q-1)$, then $r \mid q^{k-2}(q-1)$, and hence $\rho^{q^{k-2}(q-1)} \equiv 1 \ (\mathrm{MOD}\ \pi^k)$, contrary to our earlier condition on ρ. Thus $r = q^{k-1}(q-1) = \Phi(\pi^k)$, and ρ is a primitive root modulo π^k for each $k \geq 1$.

Now suppose that $q \equiv -1$ (mod 3), ρ_1 is a primitive root modulo 2, and ρ_2 is a primitive root modulo π^k. By the Chinese remainder theorem, there exists a ρ such that $\rho \equiv \rho_1$ (MOD 2) and $\rho \equiv \rho_2$ (MOD π^k). Thus ρ is a primitive root of both 2 and π^k. Also, since $q \equiv -1$ (mod 3), then $\gcd(\Phi(2), \Phi(\pi^k)) = \gcd(3, q^{k-1}(q-1)) = 1$, and so ρ is a primitive root of $2\pi^k$, by Lemma 7.15. Q.E.D.

LEMMA 7.22 *Let π be a prime such that $|N(\pi)| = q$, where q is a rational prime of the form $q \equiv \pm 1$ (mod 5). Also, let p be a rational prime of the form $p \equiv \pm 2$ (mod 5), $k \geq 1$, and ϵ a unit. Then the following integers do* not *have primitive roots*:
(i) ϵp^k, *whenever* $k \geq 2$;
(ii) $(2\epsilon)p^k$, *whenever* $k \geq 2$;
(iii) $\epsilon(2 + \omega)^k$, *whenever* $k \geq 3$;
(iv) $(2\epsilon)(2 + \omega)^k$, *whenever* $k \geq 3$;
(v) $(2\epsilon)\pi^k$, *whenever* $k \geq 1$ *and* $q \equiv 1$ (mod 3);
(vi) $(2\epsilon)p$, *whenever* $p \neq 3$.
PROOF: It follows from Lemma 7.19 that $\operatorname{ord}(\alpha, \epsilon p^k) < \Phi(\epsilon p^k)$ for all $k \geq 2$ and all α such that $\operatorname{GCD}(\alpha, \epsilon p^k) = 1$. Thus ϵp^k does not have a primitive root when $k \geq 2$. By Lemma 7.15, it is also apparent that $(2\epsilon)p^k$ does not have a primitive root when $k \geq 2$. This establishes (i) and (ii); and (iii) and (iv) are established in a similar fashion with the aid of Lemma 7.18. Turning our attention to (v), we observe that if $q \equiv 1$ (mod 3), then 3 divides $q^{k-1}(q-1) = \Phi(\pi^k)$, so that $\gcd(\Phi(2\epsilon), \Phi(\pi^k)) = 3$. Consequently, if $q \equiv 1$ (mod 3), then it follows from Lemma 7.15 that $(2\epsilon)\pi^k$ does not have a primitive root. Thus (v) is established. Finally, if $p \neq 3$, then 3 divides $(p-1)(p+1) = \Phi(p)$, and so $\gcd(\Phi(2\epsilon), \Phi(p)) = 3$. Thus, if $p \neq 3$, it follows from the familiar Lemma 7.15 that $(2\epsilon)p$ does not have a primitive root, and the proof is complete. Q.E.D.

THEOREM 7.23 (Integers with primitive roots) *Let π be a prime of the form $|N(\pi)| = q$, where q is a rational prime of the form $q \equiv \pm 1$ (mod 5), and let p be a rational prime of the form $p \equiv \pm 2$ (mod 5). Then the following integers have primitive roots:*

(i) p;

(ii) π^k, *whenever* $k \geq 1$;

(iii) $2\pi^k$, *whenever* $k \geq 1$ *and* $q \equiv -1 \pmod 3$;

(iv) 1;

(v) 6;

(vi) $2 + \omega$, $(2 + \omega)^2$, $2(2 + \omega)$, *and* $2(2 + \omega)^2$.

Moreover, μ has a primitive root if and only if μ is an associate of one of the integers given in (i)–(vi).

PROOF: It follows from Lemmas 7.17 and 7.22 that associates of the integers given in (i)–(vi) are the only possible integers having primitive roots. Thus it suffices to show that the integers in (i)–(vi) have primitive roots. By Corollary 7.11 and Lemma 7.21, we deduce that the integers in (i)–(iii) have primitive roots. The integer 1 is a primitive root modulo 1, and it is easily verified that ω is a primitive root modulo 6. Thus the proof is complete when we find primitive roots for the integers in (vi). We shall in fact show that ω is a primitive root for each of the integers in (vi). By direct computation, we see that ω is a primitive root for both $2 + \omega$ and 5; and since $(2 + \omega)^2$ is an associate of 5, ω is a primitive root for $(2 + \omega)^2$. Finally, upon observing that ω is a primitive root of 2, an easy application of Lemma 7.15 shows that ω is a primitive root of $2(2 + \omega)$ and $2(2 + \omega)^2$. Q.E.D.

The following theorem gives the number of incongruent primitive roots of an integer μ and, if one primitive root is known, how to calculate the remaining ones.

THEOREM 7.24 *If an integer μ has a primitive root ρ, then μ has $\phi(\Phi(\mu))$ incongruent primitive roots modulo μ and these are given by the integers ρ^k such that $0 \leq k \leq \Phi(\mu) - 1$ and $\gcd(k, \Phi(\mu)) = 1$.*

PROOF: By Corollary 7.9, $\{1, \rho, \rho^2, \ldots, \rho^{\Phi(\mu)-1}\}$ is a reduced set of residues modulo μ. Thus every integer relatively prime to μ is congruent modulo μ to a unique integer in this set. In particular, every primitive root of μ is congruent modulo μ to one of the integers in this set. Thus we need only determine the rational integers k in the range $0 \leq k \leq \Phi(\mu) - 1$ such that $\text{ord}(\rho^k, \mu) = \Phi(\mu) = \text{ord}(\rho, \mu)$. It follows from Theorem 7.6 that $\text{ord}(\rho^k, \mu) = \Phi(\mu)$ if and only if $1 = \text{ord}(k, \Phi(\mu))$. Since there are exactly $\phi(\Phi(\mu))$ rational integers k such that $0 \leq k \leq \Phi(\mu) - 1$ and $\gcd(k, \Phi(\mu)) = 1$, the theorem is proved. Q.E.D.

Indices

DEFINITION Let ρ be a primitive root modulo μ and GCD(α, μ) = 1. The unique rational integer k satisfying $0 \leq k \leq \Phi(\mu) - 1$ and $\alpha \equiv \rho^k$ (MOD μ) is called the **index of α to the base ρ modulo μ,** and we write $k = \text{ind}_\rho(\alpha)$ or simply $k = \text{ind}(\alpha)$ when the base need not be specifically mentioned. Thus $k = \text{ind}_\rho(\alpha)$ if and only if $\alpha \equiv \rho^k$ (MOD μ) and $0 \leq k \leq \Phi(\mu) - 1$.

The theory of indices for $\hat{Z}$ is developed along the same lines as in rational number theory. Indices are useful in that they may be used to reduce certain algebraic congruences modulo μ to an equivalent rational integer congruence modulo $\Phi(\mu)$. The following theorem gives the basic properties of the index function.

THEOREM 7.25 *If ρ is a primitive root modulo μ, GCD(α, μ) = GCD(β, μ) = 1, and $n \geq 0$, then*
(i) $\alpha \equiv \beta$ (MOD μ) *if and only if* $\text{ind}_\rho(\alpha) = \text{ind}_\rho(\beta)$;
(ii) $\text{ind}_\rho(\alpha\beta) \equiv \text{ind}_\rho(\alpha) + \text{ind}_\rho(\beta)$ (mod $\Phi(\mu)$);
(iii) $\text{ind}_\rho(\alpha^n) \equiv n \cdot \text{ind}_\rho(\alpha)$ (mod $\Phi(\mu)$);
(iv) $\text{ind}_\rho(\alpha) \equiv \text{ind}_{\rho_1}(\alpha) \cdot \text{ind}_\rho(\rho_1)$ (mod $\Phi(\mu)$), *where ρ_1 is also a primitive root modulo μ;*
(v) $\text{ind}_\rho(1) = 0$;
(vi) $\text{ind}_\rho(-1) = \frac{1}{2}\Phi(\mu)$, *if μ is not a unit nor an associate of* 2.
PROOF: The proofs of (i), (ii), and (iii) are established exactly as in rational integer theory and will not be repeated here. In order to verify (iv), set $r = \text{ind}_\rho(\alpha)$, $s = \text{ind}_{\rho_1}(\alpha)$, and $t = \text{ind}_\rho(\rho_1)$. Then we have $\alpha \equiv \rho^r$ (MOD μ), $\alpha \equiv \rho_1^s$ (MOD μ), and $\rho_1 \equiv \rho^t$ (MOD μ). This gives us $\rho^r \equiv \alpha \equiv \rho_1^s \equiv (\rho^t)^s = \rho^{st}$ (MOD μ), and, as a consequence, $r \equiv st$ (mod $\Phi(\mu)$), by Theorem 7.2(iv). This is the desired relationship (iv). Assertion (v) is trivial. To establish (vi), we must show that $\rho^{\Phi(\mu)/2} \equiv -1$ (MOD μ) whenever μ is not a unit nor an associate of 2. By Theorem 7.23, it follows that $\mu = \epsilon\pi^k$ or $\mu = (2\epsilon) \cdot \pi^k$, where $k \geq 1$, ϵ is a unit, and the prime π is not an associate of 2. For these values of μ it follows, by Lemma 7.16, that $\Phi(\mu)$ is an even rational integer, so that

$$\left(\rho^{\Phi(\mu)/2} - 1\right)\left(\rho^{\Phi(\mu)/2} + 1\right) = \rho^{\Phi(\mu)} - 1 \equiv 0 \ (\text{MOD } \mu).$$

Consequently,

$$\mu \mid \left(\rho^{\Phi(\mu)/2} - 1\right)\left(\rho^{\Phi(\mu)/2} + 1\right). \tag{7.2}$$

We next show that $\pi^k \mid (\rho^{\Phi(\mu)/2} + 1)$ via an argument by contradiction. Consider the equation

$$\left(\rho^{\Phi(\mu)/2} + 1\right) - \left(\rho^{\Phi(\mu)/2} - 1\right) = 2. \tag{7.3}$$

If $\pi^k \nmid (\rho^{\Phi(\mu)/2} + 1)$, it would follow from (7.2) that $\pi \mid (\rho^{\Phi(\mu)/2} - 1)$. But since 2 is a prime and π is not an associate of 2, then $\pi \nmid 2$, and so it would follow from (7.3) that $\pi \nmid (\rho^{\Phi(\mu)/2} + 1)$. Hence, by (7.2), we would conclude that $\pi^k \mid (\rho^{\Phi(\mu)/2} - 1)$ and, consequently,

$$\rho^{\Phi(\mu)/2} \equiv 1 \ (\text{MOD } \pi^k). \tag{7.4}$$

If $\mu = \epsilon\pi^k$, then (7.4) is an obvious contradiction to the fact that ρ is a primitive root of μ. On the other hand, if $\mu = (2\epsilon)\pi^k$, then ρ would be a primitive root modulo π^k, by Lemma 7.15. It would then follow from (7.4) that $\Phi(\pi^k)$ divides $\frac{1}{2}\Phi(2\epsilon\pi^k) = \frac{1}{2}(3)\Phi(\pi^k)$, which is impossible. Thus the assumption that $\pi^k \nmid (\rho^{\Phi(\mu)/2} + 1)$ is false, and so we deduce that $\rho^{\Phi(\mu)/2} \equiv -1 \ (\text{MOD } \pi^k)$. Accordingly, if $\mu = \epsilon\pi^k$, we are done. If $\mu = (2\epsilon)\pi^k$, then 2, being a prime, must divide one of the two factors of (7.2). But then 2 must divide both factors, because of (7.3). Thus we have both $\rho^{\Phi(\mu)/2} \equiv -1$ (MOD 2) and $\rho^{\Phi(\mu)/2} \equiv -1 \ (\text{MOD } \pi^k)$. Consequently, $\rho^{\Phi(\mu)/2} \equiv -1 \ (\text{MOD } \mu)$, and the proof is complete. Q.E.D.

Using the primitive root ω of 6, we construct the following index table.

ξ	$\text{ind}(\xi)$	ξ	$\text{ind}(\xi)$	ξ	$\text{ind}(\xi)$
1	0	$1 + 2\omega$	3	$3 + 4\omega$	9
5	12	$3 + 2\omega$	21	$5 + 4\omega$	15
ω	1	$5 + 2\omega$	6	5ω	13
$1 + \omega$	2	$1 + 3\omega$	8	$1 + 5\omega$	11
$2 + \omega$	7	$2 + 3\omega$	4	$2 + 5\omega$	22
$3 + \omega$	17	$4 + 3\omega$	16	$3 + 5\omega$	5
$4 + \omega$	10	$5 + 3\omega$	20	$4 + 5\omega$	19
$5 + \omega$	23	$1 + 4\omega$	18	$5 + 5\omega$	14

In the next example we illustrate the use of indices by solving several algebraic congruences with modulus 6.

EXAMPLE 7.5 The algebraic congruence $(2 + \omega)X \equiv 5$ (MOD 6) is equivalent to the algebraic congruence

$$\text{ind}(2 + \omega) + \text{ind}(X) \equiv \text{ind}(5) \text{ (mod 24)}.$$

Using the index table and simplifying the last congruence, we obtain $\text{ind}(X) \equiv 5$ (mod 24). Consequently, $\text{ind}(X) = 5$ and, using the table, we conclude that $3 + 5\omega$ is the solution of the congruence $(2 + \omega)X \equiv 5$ (MOD 6).

The algebraic congruence $(4 + \omega)X^9 \equiv \omega$ (MOD 6) is equivalent to $\text{ind}(4 + \omega) + 9\text{ind}(X) \equiv \text{ind}(\omega)$ (mod 24). Using the index table and simplifying, we obtain the linear congruence $9\text{ind}(X) \equiv -9$ (mod 24), which is equivalent to $\text{ind}(X) \equiv -1$ (mod 8). Consequently, $\text{ind}(X) = 7$, 15, or 23. Thus $2 + \omega$, $5 + 4\omega$, and $5 + \omega$ are the three solutions of the congruence $(4 + \omega)X^9 \equiv \omega$ (MOD 6).

VIII   Quadratic
Residues

Introduction

An integer α is said to be a *quadratic residue* or *quadratic nonresidue* of a prime π according as the congruence $X^2 \equiv \alpha$ (MOD π) has or has not a solution. The major accomplishment of this chapter is Theorem 8.5. This theorem gives a computationally efficient procedure to determine whether α is a quadratic residue or nonresidue of π. This determination of the quadratic nature of an integer is important in the study of quadratic congruences, for we have the following fundamental result: If the prime π is not an associate of 2 and $GCD(\alpha, \pi) = 1$, then the quadratic congruence

$$\alpha X^2 + \beta X + \gamma \equiv 0 \text{ (MOD } \pi) \qquad (8.1)$$

has a solution if and only if the integer $\delta = \beta^2 - 4\alpha\gamma$ is a quadratic residue of π. We shall verify this result in the next paragraph. (When π is an associate of 2, the solvability of (8.1) is ascertained by substitution of the integers 0, 1, ω, and ω^2 into the congruence.) Hence, in the case of quadratic congruences, Theorem 8.5 will resolve one of the important problems left open in Chapter 6. For recall, in that chapter no practical procedure for determining the solvability of the congruence (8.1) was given. Indeed, the procedure used was the highly inefficient method of substituting the $|N(\pi)|$ integers of a complete set of residues into the congruence.

In order to verify the assertion of the previous paragraph, we note that since π is not an associate of 2 and $GCD(\alpha, \pi) = 1$, then $GCD(4\alpha, \pi) = 1$. This means that (8.1) is equivalent to the congruence $4\alpha(\alpha X^2 + \beta X + \gamma) \equiv 0$ (MOD π), and since $4\alpha(\alpha X^2 + \beta X + \gamma) = (2\alpha X + \beta)^2 - (\beta^2 - 4\alpha\gamma)$, then (8.1) is seen to be equivalent to

$$(2\alpha X + \beta)^2 \equiv \beta^2 - 4\alpha\gamma \text{ (MOD } \pi).$$

If we now set $Y = 2\alpha X + \beta$, and $\delta = \beta^2 - 4\alpha\gamma$, we obtain

$$Y^2 \equiv \delta \text{ (MOD } \pi). \tag{8.2}$$

If ξ is a solution of (8.1), then $\eta = 2\alpha\xi + \beta$ is a solution of (8.2). Conversely, if η is a solution of (8.2), then we can solve the linear congruence $2\alpha X + \beta \equiv \eta$ (MOD π) and obtain a solution ξ of (8.1). Hence we see that (8.1) has a solution if and only if $\delta = \beta^2 - 4\alpha\gamma$ is a quadratic residue of π. It should be noted that if the solutions of (8.2) are known, then the foregoing discussion shows how these solutions may be used to solve congruence (8.1).

The theory of this chapter is closely related to quadratic residue theory for rational integers. The latter theory will be assumed. In particular, we shall assume the properties of the Legendre symbol $(a \mid p)$, and its evaluation by way of the celebrated reciprocity law of Gauss. We begin the development with the more general concept of an nth power residue and prove the standard theorem for these residues. The remainder of the chapter is devoted to the specific case of quadratic residues when the modulus is a prime.

Power Residues

DEFINITIONS   If $n > 0$, then α is said to be an **nth power residue modulo μ** (or **of μ**) if the algebraic congruence $X^n \equiv \alpha$ (MOD μ)

has at least one solution. We shall write $\alpha R_n \mu$ or $\alpha N_n \mu$ according as α is or is not an nth power residue modulo μ. In the case $\alpha R_2 \mu$, we say that α is a **quadratic residue of μ**; if $\alpha N_2 \omega$ we say that α is a **quadratic nonresidue of μ**. Similarly, when $n = 3$ we speak of **cubic residues** and **cubic nonresidues**.

EXAMPLE 8.1 Taking $\{0, 1, \omega, \omega^2\}$ as a complete set of residues modulo 2, we have the following table for the squares and cubes of integers modulo 2.

ξ	ξ^2	ξ^3
0	0	0
1	1	1
ω	ω^2	1
ω^2	ω	1

From this table we observe that every integer is a quadratic residue of 2, and that an integer α is a cubic residue of 2 if and only if α is 0 or 1 modulo 2. As we shall shortly see, associates of the prime 2 are the only primes π such that $\alpha R_2 \pi$ for every α.

For the remainder of the chapter we shall assume that the modulus μ is a prime π. Also, if π is a prime and $\alpha \equiv 0 \ (\text{MOD } \pi)$, then $X^n \equiv \alpha \ (\text{MOD } \pi)$ has the solution α for every $n > 0$; that is, $\alpha R_n \pi$ whenever $\text{GCD}(\alpha, \pi) = \pi$. Thus we shall dispense with this uninteresting case and only consider the integers α that are relatively prime to π. We have the following basic theorem on nth power residues.

THEOREM 8.1 Let π be a given prime and $n > 0$. If $\text{GCD}(\alpha, \pi) = 1$ and $d = \gcd(n, \Phi(\pi))$, then
(i) $\alpha R_n \pi$ if and only if $\alpha^{\Phi(\pi)/d} \equiv 1 \ (\text{MOD } \pi)$;
(ii) if $\alpha R_n \pi$, the algebraic congruence

$$X^n \equiv \alpha \ (\text{MOD } \pi) \tag{8.3}$$

has exactly d incongruent solutions;
(iii) there are exactly $\Phi(\pi)/d$ incongruent nth power residues modulo π that are relatively prime to π.
PROOF: Let ρ be a primitive root modulo π. Since α is relatively

prime to π, so is any solution ξ of (8.3). Consequently, $\mathrm{ind}_\rho(\xi)$, for any such solution, does exist. Thus it follows from the theory of indices that (8.3) is equivalent to the rational integer congruence

$$n \cdot \mathrm{ind}_\rho(X) \equiv \mathrm{ind}_\rho(\alpha) \ (\mathrm{mod}\ \Phi(\pi)). \tag{8.4}$$

By a standard result in rational integer theory, this linear congruence in $\mathrm{ind}_\rho(X)$ has a rational integer solution if and only if $d \mid \mathrm{ind}_\rho(\alpha)$. It is easy to see that $d \mid \mathrm{ind}_\rho(\alpha)$ if and only if $d^{-1} \cdot \Phi(\pi) \cdot \mathrm{ind}_\rho(\alpha) \equiv 0 \ (\mathrm{mod}\ \Phi(\pi))$, and since this congruence is equivalent to $\alpha^{\Phi(\pi)/d} \equiv 1 \ (\mathrm{MOD}\ \pi)$, then (8.3) has a solution if and only if $\alpha^{\Phi(\pi)/d} \equiv 1 \ (\mathrm{MOD}\ \pi)$. This establishes (i).

Again from rational integer theory, if (8.4) has a rational integer solution, then it has exactly $d = \gcd(n, \Phi(\pi))$ incongruent rational integer solutions. Hence (ii) is true.

Finally, in establishing (i) we showed that (8.4) has a solution if and only if $d \mid \mathrm{ind}_\rho(\alpha)$. The values of $\mathrm{ind}_\rho(\alpha)$ such that $d \mid \mathrm{ind}_\rho(\alpha)$ are $\Phi(\pi)/d$ in number and are given by $0, d, 2d, 3d, \ldots, (d^{-1} \cdot \Phi(\pi) - 1)d$. Accordingly, the $\Phi(\pi)/d$ integers α given by $1, \rho^d, \rho^{2d}, \rho^{3d}, \ldots, \rho^{\Phi(\pi)-d}$ are the incongruent nth power residues modulo π that are relatively prime to π. Thus the result (iii) is true, and the theorem is proved. Q.E.D.

EXAMPLE 8.2 We have $3 = \gcd(9, 48) = \gcd(9, \Phi(7))$, and so, by Theorem 8.1(iii), there are $\Phi(7)/3 = 48/3 = 16$ ninth power residues of 7 that are relatively prime to 7. Since $2 + \omega$ is a primitive root modulo 7, it follows from Theorem 8.1(i) that these 16 ninth power residues are given by $1, (2 + \omega)^3, (2 + \omega)^6, (2 + \omega)^9, \ldots, (2 + \omega)^{45}$.

In Example 8.1 we saw that every integer is a quadratic residue of π whenever π is an associate of 2. For primes π that are not associates of 2, we have the following theorem.

THEOREM 8.2 *Let π be a prime which is not an associate of* 2 *and α an integer such that* $\mathrm{GCD}(\alpha, \pi) = 1$. *Then*
(i)

$$\alpha^{\Phi(\pi)/2} \equiv \begin{cases} 1 \ (\mathrm{MOD}\ \pi), & \text{if } \alpha R_2 \pi \\ -1 \ (\mathrm{MOD}\ \pi), & \text{if } \alpha N_2 \pi; \end{cases}$$

(ii) *if* $\alpha R_2 \pi$, *the algebraic congruence* $X^2 \equiv \alpha$ (MOD π) *has just two solutions modulo* π, *these being of the form* $\pm \xi$ *for some integer* ξ *relatively prime to* π;

(iii) *there are* $\frac{1}{2} \Phi(\pi)$ *incongruent quadratic residues and* $\frac{1}{2} \Phi(\pi)$ *incongruent quadradic nonresidues modulo* π *that are relatively prime to* π.

PROOF: Set $n = 2$ in Theorem 8.1. Then, since $\Phi(\pi)$ is even, we have $2 = \gcd(2, \Phi(\pi))$. This establishes the first part of (i) as well as (iii), and the assertion concerning the number of solutions in (ii). Hence all that remains is to establish the second part of (i) and the form of the solutions in (ii). The form $\pm \xi$ of the solutions in (ii) follows at once, for if $\xi^2 \equiv \alpha$ (MOD π), then $\xi \not\equiv -\xi$ (MOD π) and $(-\xi)^2 = \xi^2 \equiv \alpha$ (MOD π). To prove the second part of (i), we note that since $\alpha^{\Phi(\pi)} - 1 \equiv 0$ (MOD π) and $\alpha^{\Phi(\pi)} - 1 = (\alpha^{\Phi(\pi)/2} - 1) \cdot (\alpha^{\Phi(\pi)/2} + 1)$, then either $\alpha^{\Phi(\pi)/2} \equiv 1$ (MOD π) or $\alpha^{\Phi(\pi)/2} \equiv -1$ (MOD π). Now, by Theorem 8.1(i) applied to $n = 2$, α is a quadratic nonresidue of π if and only if $\alpha^{\Phi(\pi)/2} \not\equiv 1$ (MOD π). Consequently, α is a quadratic nonresidue of π if and only if $\alpha^{\Phi(\pi)/2} \equiv -1$ (MOD π). Q.E.D.

Just as in rational integer theory, (i) of Theorem 8.2 is referred to as *Euler's criterion for quadratic residues.* It is an important criterion for theoretical purposes but is not a practical method of determining whether a given integer is or is not a quadratic residue of a prime π (unless $\Phi(\pi)$ is small). For example, in order to decide whether $17 + 27\omega$ is a quadratic residue of 557, we would have to determine whether $(17 + 27\omega)^{155124}$ is equal to 1 or to -1 modulo 557.

COROLLARY 8.3 *Let* ρ *be a primitive root of a prime* π *which is not an associate of* 2. *Then* $1, \rho^2, \rho^4, \ldots, \rho^{\Phi(\pi)-2}$ *are the incongruent quadratic residues of* π *that are relatively prime to* π, *and* $\rho, \rho^3, \rho^5, \ldots, \rho^{\Phi(\pi)-1}$ *are the incongruent quadratic nonresidues of* π *that are relatively prime to* π.

PROOF: By Theorem 8.2(iii), there are $\frac{1}{2} \Phi(\pi)$ incongruent quadratic residues modulo π that are relatively prime to π, and it is clear that each of the $\frac{1}{2} \Phi(\pi)$ incongruent integers $1, \rho^2, \rho^4, \ldots, \rho^{\Phi(\pi)-2}$ is such a quadratic residue. This means that the $\frac{1}{2} \Phi(\pi)$ incongruent integers $\rho, \rho^3, \rho^5, \ldots, \rho^{\Phi(\pi)-1}$ are the quadratic nonresidues of π that are relatively prime to π. Q.E.D.

Legendre Symbol for $\hat{Z}$

Analogous to rational integer theory, it is convenient to define a Legendre symbol for $\hat{Z}$ to aid in determining the quadratic character of an integer. It is important that the notation used to denote the Legendre symbol for $\hat{Z}$ be such that no confusion with the Legendre symbol $(\alpha \mid p)$ for Z will ensue, for we shall eventually evaluate the Legrendre symbol for $\hat{Z}$ in terms of the Legendre symbol for Z. With these comments in mind, we make the following definition.

DEFINITION Let π be a prime which is not an associate of 2 and $\mathrm{GCD}(\alpha, \pi) = 1$. The Legendre symbol $[\alpha \mid \pi]$ is defined by

$$\left[\,\alpha \mid \pi\,\right] = \left\{ \begin{array}{ll} 1, & \text{if } \alpha R_2 \pi \\ -1, & \text{if } \alpha N_2 \pi. \end{array} \right.$$

Thus the problem of determining whether α is a quadratic residue or nonresidue of π now becomes the problem of finding the value of $[\alpha \mid \pi]$. The following theorem shows that the Legendre symbol $[\alpha \mid \pi]$ has the same basic properties as the symbol $(\alpha \mid p)$.

THEOREM 8.4 *Let* π *be a prime which is not an associate of* 2. *If* $\mathrm{GCD}(\alpha, \pi) = \mathrm{GCD}(\beta, \pi) = 1$, *then we have the following*:
(i) $[\alpha \mid \pi] \equiv \alpha^{\Phi(\pi)/2} \ (\mathrm{MOD} \ \pi)$.
(ii) $[\alpha \mid \pi] \cdot [\beta \mid \pi] = [\alpha\beta \mid \pi]$.
(iii) *If* $\alpha \equiv \beta \ (\mathrm{MOD} \ \pi)$, *then* $[\alpha \mid \pi] = [\beta \mid \pi]$.
(iv) *If* π_1 *is an associate of* π, *then* $[\alpha \mid \pi_1] = [\alpha \mid \pi]$.
(v) *If* $\mathrm{GCD}(\gamma, \pi) = 1$, *then* $[\alpha\gamma^2 \mid \pi] = [\alpha \mid \pi]$.
(vi) $[1 \mid \pi] = 1$ *and* $[-1 \mid \pi] = (-1)^{\Phi(\pi)/2}$.
PROOF: Assertion (i) follows immediately from the definition of Legendre symbol and Theorem 8.2(i). Using (i) of this theorem, we obtain

$$\left[\,\alpha \mid \pi\,\right] \cdot \left[\,\beta \mid \pi\,\right] \equiv \alpha^{\Phi(\pi)/2} \cdot \beta^{\Phi(\pi)/2} = (\alpha\beta)^{\Phi(\pi)/2} \equiv \left[\,\alpha\beta \mid \pi\,\right] \ (\mathrm{MOD} \ \pi).$$

The result (ii) now follows, since each of the expressions $[\alpha \mid \pi] \cdot [\beta \mid \pi]$ and $[\alpha\beta \mid \pi]$ takes the value -1 or 1 and $-1 \not\equiv 1 \ (\mathrm{MOD} \ \pi)$.

Assertion (iii) follows by a similar argument, since if $\alpha \equiv \beta$ (MOD π), then

$$[\alpha \,|\, \pi] \equiv \alpha^{\Phi(\pi)/2} \equiv \beta^{\Phi(\pi)/2} \equiv [\beta \,|\, \pi] \text{ (MOD } \pi).$$

Assertion (iv) follows at once upon noting that ξ satisfies the congruence $X^2 \equiv \alpha$ (MOD π) if and only if ξ satisfies the congruence $X^2 \equiv \alpha$ (MOD π_1). To establish (v) we need only apply (ii) and observe that $[\gamma^2 \,|\, \pi] = 1$. The first part of (vi) is trivial, and the second part follows from (i) upon setting $\alpha = -1$, and arguing equality as in the proof of (ii). Q.E.D.

Evaluation of $[\alpha \,|\, \pi]$

As earlier noted, the following theorem is the most important result of this chapter, for it gives an efficient method for evaluating the symbol $[\alpha \,|\, \pi]$. Of course, Theorem 8.4 has already handled the special cases $\alpha = 1$ and $\alpha = -1$.

THEOREM 8.5 *Let p be an odd rational prime of the form $p \equiv \pm 2$ (mod 5) and π a prime such that $|\mathrm{N}(\pi)| = q$, where q is a rational prime (necessarily odd). Then we have the following:*
(i) *If $\mathrm{GCD}(\alpha, p) = 1$, then $[\alpha \,|\, p] = (\mathrm{N}(\alpha) \,|\, p)$.*
(ii) *If $\mathrm{GCD}(a, \pi) = 1$, then $[a \,|\, \pi] = (a \,|\, q)$.*
PROOF: Using Theorem 5.51(i), Theorem 8.4(i), and the result $(\mathrm{N}(\alpha))^{(p-1)/2} \equiv (\mathrm{N}(\alpha) \,|\, p)$ (mod p) from rational integer theory, we obtain that $[\alpha \,|\, p] \equiv \alpha^{\Phi(p)/2} = (\alpha^{p+1})^{(p-1)/2} \equiv (\mathrm{N}(\alpha))^{(p-1)/2} \equiv (\mathrm{N}(\alpha) \,|\, p)$ (MOD p). Hence $[\alpha \,|\, p] \equiv (\mathrm{N}(\alpha) \,|\, p)$ (MOD p), and since each side is either 1 or -1 and $-1 \not\equiv 1$ (MOD p), then we have $[\alpha \,|\, p] = (\mathrm{N}(\alpha) \,|\, p)$. Thus (i) is true.

In order to establish (ii), we first note, by Theorem 5.16, that $a^{(q-1)/2} \equiv b$ (MOD π) if and only if $a^{(q-1)/2} \equiv b$ (mod q). Consequently, since $a^{(q-1)/2} \equiv (a \,|\, q)$ (mod q) from rational integer theory, then $[a \,|\, \pi] \equiv a^{\Phi(\pi)/2} = a^{(q-1)/2} \equiv (a \,|\, q)$ (mod q). Thus $[a \,|\, \pi]$

$\equiv (a \,|\, q)$ (mod q) and it follows, by the same reasoning used in (i), that $[a \,|\, \pi] = (a \,|\, q)$. Hence (ii) is true, and the theorem is established. Q.E.D.

A few elementary but nevertheless important observations regarding the utility of Theorem 8.5 need to be emphasized. First, if π is a prime which is not an associate of 2, then π is either an associate of an odd rational prime p of the form $p \equiv \pm 2$ (mod 5), or else there exists an odd rational prime q such that $|\mathrm{N}(\pi)| = q$. In the former case, if $\mathrm{GCD}(\alpha, \pi) = 1$, then we have $[\alpha \,|\, \pi] = [\alpha \,|\, p]$ $= (\mathrm{N}(\alpha) \,|\, p)$. In the latter case, if $\mathrm{GCD}(\alpha, \pi) = 1$, then since $\{1, 2, 3, \ldots, q - 1\}$ is a reduced set of residues modulo π, there exists a unique rational integer a in the range $1 \leq a \leq q - 1$ such that $\alpha \equiv a$ (MOD π). Consequently, in this case we have $[\alpha \,|\, \pi] = [a \,|\, \pi]$ $= (a \,|\, q)$. The rational integer a may be determined in the manner of Example 5.3. Thus, if π is any prime which is not an associate of 2 and $\mathrm{GCD}(\alpha, \pi) = 1$, then we may readily convert the evaluation of $[\alpha \,|\, \pi]$ into a problem of evaluating a Legendre symbol $(a \,|\, p)$. This is a happy state of affairs, for it allows us to use rational integer theory, in particular, the quadratic reciprocity law of Gauss, to quickly evaluate $[\alpha \,|\, \pi]$. We illustrate these ideas in the next two examples.

EXAMPLE 8.3 557 is a rational prime, $557 \equiv 2$ (mod 5), $\mathrm{GCD}(17 + 27\omega, 557) = 1$, and $\mathrm{N}(17 + 27\omega) = 19$. Therefore, by Theorem 8.5(i), we obtain $[17 + 27\omega \,|\, 557] = (\mathrm{N}(17 + 27\omega) \,|\, 557)$ $= (19 \,|\, 557)$. Evaluating $(19 \,|\, 557)$, we obtain $(19 \,|\, 557) = 1$. Thus $[17 + 27\omega \,|\, 557] = 1$, and so $17 + 27\omega$ is a quadratic residue of 557.

EXAMPLE 8.4 We have $|\mathrm{N}(20 + 7\omega)| = 491$, $|\mathrm{N}(3 + 11\omega)| = 79$, and both 79 and 491 are rational primes. Therefore $20 + 7\omega$ and $3 + 11\omega$ are nonassociated primes. We wish to evaluate the Legendre symbols $[20 + 7\omega \,|\, 3 + 11\omega]$ and $[3 + 11\omega \,|\, 20 + 7\omega]$. Using the method of Example 5.3, we obtain $20 + 7\omega \equiv 54$ (MOD $3 + 11\omega$) and $3 + 11\omega \equiv 182$ (MOD $20 + 7\omega$). Hence, by Theorem 8.5(ii), we have

$$\left[20 + 7\omega \,|\, 3 + 11\omega \right] = \left[54 \,|\, 3 + 11\omega \right] = (54 \,|\, 79)$$

and

$$[3 + 11\omega \mid 20 + 7\omega] = [182 \mid 20 + 7\omega] = (182 \mid 491).$$

Evaluating the Legendre symbols $(54 \mid 79)$ and $(182 \mid 491)$, we obtain $(54 \mid 79) = -1$ and $(182 \mid 491) = 1$. Consequently, $[20 + 7\omega \mid 3 + 11\omega] = -1$ and $[3 + 11\omega \mid 20 + 7\omega] = 1$.

Quadratic Character of ω

We bring this chapter to a close by determining the quadratic character of the units ω and ω^{-1}. With this information and the properties listed in Theorem 8.4, it will then be easy to determine the quadratic character of any unit $\epsilon = \pm \omega^n$.

THEOREM 8.6 *If π is a prime which is not an associate of* 2, *then* $[\omega^{-1} \mid \pi] = [\omega \mid \pi]$.
PROOF: Since $1 = [1 \mid \pi] = [\omega^{-1}\omega \mid \pi] = [\omega^{-1} \mid \pi] \cdot [\omega \mid \pi]$, then $[\omega^{-1} \mid \pi] = [\omega \mid \pi]$. Q.E.D.

THEOREM 8.7 *Let p be an odd rational prime of the form* $p \equiv \pm 2 \pmod 5$ *and $c + d\omega$ a prime such that* $|N(c + d\omega)| = q$, *where q is a rational prime. Then* $[\omega \mid p] = (-1)^{(p-1)/2}$ *and* $[\omega \mid c + d\omega] = (-cd \mid q)$.
PROOF: Since $(-1 \mid p) = (-1)^{(p-1)/2}$, then $[\omega \mid p] = (N(\omega) \mid p) = (-1 \mid p) = (-1)^{(p-1)/2}$. As noted prior to Example 5.3 in Chapter 5, $\gcd(d, q) = 1$, and so $d^{q-1} \equiv 1 \pmod q$. Consequently, in light of Theorem 5.16 and the result $d\omega \equiv -c \pmod{c + d\omega}$, we have $\omega \equiv d^{q-1}\omega \equiv -cd^{q-2} \pmod{c + d\omega}$. Since $q - 3$ is an even rational integer, then $(d^{q-3} \mid q) = 1$. Hence, by Theorem 8.5(ii), $[\omega \mid c + d\omega] = [-cd^{q-2} \mid c + d\omega] = (-cd^{q-2} \mid q) = (d^{q-3} \mid q) \cdot (-cd \mid q) = (-cd \mid q)$. Q.E.D.

IX   Applications to Rational Integer Theory

Introduction

In Chapter 3 we solved the Diophantine equation $x^2 + xy - y^2 = M$ with the aid of number theory in $Z(\omega)$. We now consider additional applications to Mersenne numbers, Fermat's last theorem, and Fibonacci numbers.

Lucas's Test for Primeness of the Mersenne Number M_{4n+3}

If p is a rational prime, then the number $M_p = 2^p - 1$ is called a *Mersenne number*. An important problem in number theory is to determine whether a given Mersenne number M_p is a rational prime. In 1876 Edouard Lucas (see Dickson [6], vol. 1, p. 22) devised a test which gave necessary and sufficient conditions for

M_p to be a rational prime whenever p is a rational prime of the form $p \equiv 3 \pmod 4$. He used this test to verify that M_{67} is composite and M_{127} is a rational prime. Lucas never published a complete proof of this test, but numerous proofs have since been given.

In this section we state the Lucas test (Theorem 9.2) and give a proof which is due to Western (Western [22]). The same proof may also be found in Hardy and Wright [9], pp.223–225.

NOTATION If $n > 0$, then $r_n = \omega^{2^n} + \bar{\omega}^{2^n}$.

We will need the following simple result when applying the Lucas test.

THEOREM 9.1
(i) $r_1 = 3$.
(ii) $r_{n+1} = r_n^2 - 2$.
PROOF: We have $r_1 = \omega^2 + \bar{\omega}^2 = \omega + 1 + \bar{\omega} + 1 = (\omega + \bar{\omega}) + 2$
$= 3$. Also $r_n^2 - 2 = (\omega^{2^n} + \bar{\omega}^{2^n})^2 - 2 = \omega^{2^{n+1}} + 2(\omega\bar{\omega})^{2^n} + \bar{\omega}^{2^{n+1}} - 2$
$= \omega^{2^{n+1}} + 2(-1)^{2^n} + \bar{\omega}^{2^{n+1}} - 2 = r_{n+1}$. Q.E.D.

It follows from Theorem 9.1 that r_n is a positive rational integer for each $n > 0$; and $r_1 = 3$, $r_2 = 7$, $r_3 = 47$, and so on.

THEOREM 9.2 (Lucas test) *If p is a rational prime of the form $p \equiv 3 \pmod 4$ and $M_p = 2^p - 1$, then M_p is a rational prime if and only if $r_{p-1} \equiv 0 \pmod{M_p}$.*
PROOF: Throughout this proof we let $M = M_p$ and $r = r_{p-1}$. Since $p = 4n + 3$ for some $n \geq 0$, then $M = 2^p - 1 = 2^{4n+3} - 1$
$= 8 \cdot (16)^n - 1 \equiv 3 \cdot 1 - 1 \equiv 2 \pmod 5$. In particular, this means that 5 is not a factor of M. In addition, 2 is not a factor of M, because M is odd. Thus the rational integer prime decomposition of M is of the form $M = p_1 p_2 \cdots p_s q_1 q_2 \cdots q_t$, where the p_i are odd rational primes of the form $p_i \equiv \pm 2 \pmod 5$ and the q_j are odd rational primes of the form $q_j \equiv \pm 1 \pmod 5$. Since $M \equiv 2 \pmod 5$, there is at least one p_i in the decomposition. (There may be no q_j.) With these preliminary observations and notational conventions in hand, we are ready to proceed with the more

technical details of the proof; that is, the demonstration that M is a rational prime if and only if $r \equiv 0 \pmod{M}$.

Suppose first that M is a rational prime. Since $\bar{\omega} = -\omega^{-1}$, then $\bar{\omega}^{2^{p-1}} = \omega^{-2^{p-1}}$; and so we deduce that $r = \omega^{2^{p-1}} + \bar{\omega}^{2^{p-1}} = \bar{\omega}^{2^{p-1}}(\omega^{2^{p-1}} \cdot \omega^{2^{p-1}} + 1) = \bar{\omega}^{2^{p-1}}(\omega^{2^p} + 1)$. Also, by Theorem 5.51(i), we have $\omega^{2^p} = \omega^{M+1} \equiv N(\omega) = -1 \pmod{M}$. Accordingly,

$$r = \bar{\omega}^{2^{p-1}}(\omega^{2^p} + 1) \equiv \bar{\omega}^{2^{p-1}}(-1 + 1) \equiv 0 \pmod{M},$$

and one direction of the theorem is established.

Now suppose that $r \equiv 0 \pmod{M}$. Under this assumption we shall deduce that there are no q_j's in the factorization of M, and that any p_i in the factorization of M satisfies $p_i = M$. This, of course, would force the conclusion that M is a rational prime, and the proof would be complete. We first note that, since $r \equiv 0 \pmod{M}$, then

$$\omega^{2^p} + 1 = \omega^{2^{p-1}}(\omega^{2^{p-1}} + \bar{\omega}^{2^{p-1}}) = \omega^{2^{p-1}} \cdot r \equiv 0 \pmod{M}.$$

From this congruence, we conclude that

$$\omega^{2^p} \equiv -1 \pmod{M},$$

and that

$$\omega^{2^{p+1}} = (\omega^{2^p})^2 \equiv (-1)^2 = 1 \pmod{M}.$$

If now e denotes any of the rational prime factors of M, then the preceding two congruences are true when the modulus M is replaced by e. Consequently, we have

$$\omega^{2^p} \equiv -1 \pmod{e} \tag{9.1}$$

$$\omega^{2^{p+1}} \equiv 1 \pmod{e} \tag{9.2}$$

We wish to conclude from (9.1) and (9.2) that $v = 2^{p+1}$, where $v = \text{ord}(\omega, e)$. Certainly, $v \mid 2^{p+1}$, because of (9.2). Suppose, to the contrary, that $v = 2^x$, where $x < p + 1$. Then, by (9.1), we would have

$$1 = (\omega^v)^{2^{p-x}} = \omega^{2^p} \equiv -1 \pmod{e}.$$

Thus $1 \equiv -1 \pmod{e}$, which is impossible (since e is not 2). Hence we conclude that if e is a rational prime factor of M, then $\text{ord}(\omega, e) = 2^{p+1}$. Now, by Theorem 5.51, we have

$$\omega^{2(p_i + 1)} = \left(\omega^{p_i + 1}\right)^2 \equiv (N(\omega))^2 = 1 \ (\text{MOD } p_i),$$

$$\omega^{q_j - 1} \equiv 1 \ (\text{MOD } q_j),$$

and so it follows that $2^{p+1} = \text{ord}(\omega, p_i)$ divides $2(p_i + 1)$ and $2^{p+1} = \text{ord}(\omega, q_j)$ divides $q_j - 1$. We must therefore have $p_i = k_i 2^p - 1$ and $q_j = h_j 2^{p+1} + 1$ for some $k_i > 0$ and $h_j > 0$. But we cannot have $q_j = h_j 2^{p+1} + 1$, for this would entail that $q_j > 2^{p+1} > 2^p - 1 = M$. Also, if $p_i = k_i 2^p - 1$ with $k_i > 1$, then $p_i \geq 2 \cdot 2^p - 1 > 2^p - 1 = M$, which is impossible. Thus $k_i = 1$, and so $p_i = 2^p - 1 = M$. Therefore, as promised, we have shown that no q_j exists in the factorization of M and that $p_i = M$. Q.E.D.

The Lucas test given by Theorem 9.2 applies only when p is a rational prime of the form $p \equiv 3 \pmod 4$. The test can be generalized so as to apply to any odd rational prime p (see Western [22]). The generalized test is almost identical to Theorem 9.2 and reads as follows: If p is an odd rational prime and $M_p = 2^p - 1$, then M_p is a rational prime if and only if $s_{p-1} \equiv 0 \pmod{M_p}$; here s_n is defined recursively by $s_1 = 4$ and $s_{n+1} = s_n^2 - 2$. Western proves the generalized test by using the integers of the quadratic field $Q(\sqrt 3)$.

The test given in Theorem 9.2 and the cited generalization have taken on added significance with the advent of the high-speed computer. For example, in 1971 Tuckerman on an IBM 360/91 found M_{19937} (a number with 6002 digits) to be a rational prime in about 35 minutes (Tuckerman [20]).

To apply the test given in Theorem 9.2, the numbers r_1, $r_2, \ldots, r_{p-1}$ are successively reduced modulo M_p by using Theorem 9.1. For example, the following table of reductions of the r_n modulo $M_{19} = 524287$ shows M_{19} to be a rational prime. (The calculations were performed with the aid of a hand-held calculator.)

n	$r_n \pmod{M_{19}}$	n	$r_n \pmod{M_{19}}$
1	3	6	-169733
2	7	7	244924
3	47	8	-104192
4	2207	9	86240
5	152264	10	-197784

n	$r_n \pmod{M_{19}}$	n	$r_n \pmod{M_{19}}$
11	-115277	15	-124288
12	208425	16	-85226
13	132664	17	-1024
14	-53409	18	0

The Equation $x^5 + y^5 + z^5 = 0$

In this section we show in Theorem 9.12 that the Diophantine equation $x^5 + y^5 + z^5 = 0$ has no solution in nonzero rational integers x, y, z. This result immediately implies Fermat's last theorem when $n = 5$. The proof we give of Theorem 9.12 is difficult, although elementary in the sense that it does not rely on the theory developed by Kummer. To establish the theorem, it is apparent that we need only show that there do not exist nonzero x, y, z such that $\gcd(x, y, z) = 1$ and $x^5 + y^5 + z^5 = 0$.

We base the proof of Theorem 9.12 on Lemmas 9.3 through 9.11. In Lemma 9.4 we deduce that if x, y, z are nonzero, $\gcd(x, y, z) = 1$, and $x^5 + y^5 + z^5 = 0$, then exactly one of x, y, z is divisible by 2 and exactly one of x, y, z is divisible by 5. Thus we have two cases to consider: The first case is when the even integer of the set $\{x, y, z\}$ is the integer divisible by 5 and the second case is when the integer of the set $\{x, y, z\}$ divisible by 5 is odd. In Lemma 9.10 we deduce that the first case is impossible, and in Lemma 9.11 we conclude that the second case is also impossible. Theorem 9.12 is then immediate. Both cases depend on the factorization property of $\hat{Z}$ given in Lemma 9.5.

A proof of Theorem 9.12 was first given in 1825 when Legendre and Dirichlet published independent proofs. It is essentially Dirichlet's proof as reconstructed by Edwards that we give (Edwards [8], pp. 65–73). We base the proofs of Lemmas 9.6 and 9.7 on Lemma 9.5, whereas Dirichlet established these two key lemmas by other means.

DEFINITION A nonzero rational integer a is said to be a **fifth power** if $a = b^5$ for some rational integer b.

LEMMA 9.3 *If a is a rational integer, then a^5 is congruent modulo 11 to 0, -1, or 1.*
PROOF: If $11 \mid a$, then $a^5 \equiv 0$ (mod 11). If $\gcd(a, 11) = 1$, then $a^{10} \equiv 1$ (mod 11), by Fermat's theorem for rational integers. Hence, since $a^{10} - 1 = (a^5 - 1)(a^5 + 1)$, then $11 \mid (a^5 - 1)$ or $11 \mid (a^5 + 1)$. We therefore have $a^5 \equiv \pm 1$ (mod 11) whenever $\gcd(a, 11) = 1$. Q.E.D.

LEMMA 9.4 *If x, y, and z are nonzero, $\gcd(x, y, z) = 1$, and $x^5 + y^5 + z^5 = 0$, then exactly one of x, y, z, is divisible by 2, and exactly one of x, y, z is divisible by 5.*
PROOF: Since $x^5 + y^5 + z^5 = 0$, then any factor common to two of the rational integers x, y, z is also a factor of the third. But $\gcd(x, y, z) = 1$, and so it follows that x, y, z are pairwise relatively prime. Therefore at most one of x, y, z is divisible by 2, and at most one of x, y, z is divisible by 5.

If x, y, z are odd, then $x^5 + y^5 + z^5 \neq 0$. This contradiction shows that one of x, y, z is divisible by 2. The proof that one of x, y, z is divisible by 5 is considerably more difficult. Suppose, to the contrary, that 5 is relatively prime to each of x, y, z. The proof will be complete when we reach a contradiction based on this assumption.

We may write $x^5 + y^5 + z^5 = 0$ in the form

$$-z^5 = (x + y)(x^4 - x^3y + x^2y^2 - xy^3 + y^4). \tag{9.3}$$

The following argument shows that the two factors on the right side of (9.3) are relatively prime: Certainly 5 is not a divisor of $x + y$, since this would imply that $5 \mid z$. If $p \neq 5$ is a rational prime and p divides both factors on the right of (9.3), then $y \equiv -x$ (mod p) and $0 \equiv 5x^4$ (mod p). Thus we have $p \mid x$ and $p \mid (x + y)$, so that $p \mid x$ and $p \mid y$. This contradicts the fact that x and y are relatively prime. We therefore conclude that the two factors on the right side of (9.3) are relatively prime.

Since the two factors on the right side of (9.3) are relatively prime, then they are both fifth powers. This allows us to conclude that

$$x + y = a_1^5, \qquad x^4 - x^3y + x^2y^2 - xy^3 + y^4 = a_2^5, \qquad z = -a_1a_2.$$
$$(9.4)$$

By similar arguments we conclude that

$$x + z = b_1^5, \qquad x^4 - x^3z + x^2z^2 - xz^3 + z^4 = b_2^5, \qquad y = -b_1b_2,$$
$$(9.5)$$

$$y + z = c_1^5, \qquad y^4 - y^3z + y^2z^2 - yz^3 + z^4 = c_2^5, \qquad x = -c_1c_2.$$
$$(9.6)$$

Since $x^5 + y^5 + z^5 = 0$, then $x^5 + y^5 + z^5 \equiv 0 \pmod{11}$. Now $\pm 1 \pm 1 \pm 1 \not\equiv 0 \pmod{11}$, and so it follows, by Lemma 9.3, that 11 divides one of x, y, z. Without loss of generality, we suppose that $11 \mid x$. It then follows, by (9.4), (9.5), and (9.6), that

$$0 \equiv 2x = (x + y) + (x + z) - (y + z) = a_1^5 + b_1^5 - c_1^5 \pmod{11}.$$

Thus, by Lemma 9.3, it follows that 11 divides one of a_1, b_1, c_1. If $11 \mid a_1$, then $11 \mid (x + y)$ because of (9.4). Thus, if $11 \mid a_1$, then $11 \mid (x + y)$ and $11 \mid x$; and so $11 \mid x$ and $11 \mid y$. This contradicts the fact that x and y are relatively prime. Therefore $11 \nmid a_1$. By a similar argument, $11 \nmid b_1$. This means that $11 \mid c_1$. If we show that this is impossible, then the proof will be complete. If $11 \mid c_1$, then, by (9.6), we have $z \equiv -y \pmod{11}$ and $5y^4 \equiv c_2^5 \pmod{11}$. But $11 \mid x$ and so it follows from (9.4) that $y^4 \equiv a_2^5 \pmod{11}$. Hence, if $11 \mid c_1$, we have $5a_2^5 \equiv c_2^5 \pmod{11}$. It then follows, by Lemma 9.3, that $a_2 \equiv c_2 \equiv 0 \pmod{11}$. However $z = -a_1a_2$ and $x = -c_1c_2$, so that 11 divides both x and z. This contradicts the fact that x and z are relatively prime, and the proof is complete. Q.E.D.

LEMMA 9.5 *If* $N(c + d\omega)$ *is a fifth power,* $\gcd(c, d) = 1$, *and* $5 \mid d$, *then there exist* a *and* b *such that* $c + d\omega = (a + b\omega)^5$.

PROOF: It follows from Corollary 3.14 that there exist distinct rational primes $q_1, q_2, \ldots, q_s$ of the form $q_i \equiv \pm 1 \pmod 5$, primes $\pi_1, \pi_2, \ldots, \pi_s$ such that $N(\pi_i) = q_i$, a unit ϵ, and nonnegative exponents $e, c_1, c_2, \ldots, c_s$ such that $c + d\omega = \epsilon(2 + \omega)^e \pi_1^{c_1} \pi_2^{c_2} \cdots \pi_s^{c_s}$. Thus $N(c + d\omega) = \pm 5^e q_1^{c_1} q_2^{c_2} \cdots q_s^{c_s}$, and since $N(c + d\omega)$ is a fifth power, the exponents $e, c_1, c_2, \ldots, c_s$ are divisible by 5. Furthermore, if $e \neq 0$, then $(2 + \omega)^2$ divides $c + d\omega$. Since 5 is an associate of $(2 + \omega)^2$, this entails that 5 divides both c

and d, contrary to the hypothesis that $\gcd(c,d) = 1$. Thus $e = 0$ and $c + d\omega$ is of the form

$$c + d\omega = \omega^n \left(\pm \pi_1^{d_1} \pi_2^{d_2} \cdots \pi_s^{d_s} \right)^5.$$

Since $n = 5r + t$, where $t = 0$, 1, 2, 3, or 4, then it is clear that $c + d\omega$ can be written in the form $c + d\omega = \omega^t (a + b\omega)^5$, where $t = 0$, 1, 2, 3, or 4. All that remains is to show that $t = 0$. Since $\omega^5 = 3 + 5\omega \equiv 3$ (MOD 5), then $(a + b\omega)^5 \equiv a^5 + b^5 \omega^5 \equiv a + b\omega^5 \equiv a + 3b$ (MOD 5). Also, $a + 3b \not\equiv 0$ (mod 5), since this would entail that $c + d\omega \equiv 0$ (MOD 5). Now if $t \neq 0$, then $c + d\omega = \omega^t (a + b\omega)^5 \equiv (F_{t-1} + F_t\omega)(a + 3b) = (a + 3b)F_{t-1} + (a + 3b) \cdot F_t \omega$ (MOD 5). Consequently, since $d \equiv 0$ (mod 5), we must have $(a + 3b)F_t \equiv 0$ (mod 5). Thus $F_t \equiv 0$ (mod 5), which is impossible for $t = 1$, 2, 3, or 4. Accordingly, $t = 0$, and the proof is complete. Q.E.D.

LEMMA 9.6 *If c and d have opposite parity, $\gcd(c,d) = 1$, $5 \mid d$, and $N(c + d\sqrt{5})$ is a fifth power, then there exist a and b such that $c + d\sqrt{5} = (a + b\sqrt{5})^5$.*
PROOF: If $c_1 = c - d$ and $d_1 = 2d$, then $\gcd(c_1, d_1) = 1$, $5 \mid d_1$, and $c + d\sqrt{5} = c + d(2\omega - 1) = c - d + (2d)\omega = c_1 + d_1\omega$. Therefore, by Lemma 9.5, there exist a_1 and b_1 such that $c + d\sqrt{5} = c_1 + d_1\omega = (a_1 + b_1\omega)^5$. Modulo 2 we have $1 \equiv c_1 + d_1\omega = (a_1 + b_1\omega)^5 \equiv (a_1 + b_1\omega)^2 \equiv a_1^2 + b_1^2\omega^2 = a_1^2 + b_1^2 + b_1^2\omega$ (MOD 2). Consequently, $b_1 \equiv 0$ (mod 2), so that $b_1 = 2b$. Setting $a = a_1 + b$, we obtain $a_1 + b_1\omega = a_1 + 2b\omega = a_1 + b(\sqrt{5} + 1) = a_1 + b + b\sqrt{5} = a + b\sqrt{5}$. Accordingly, $c + d\sqrt{5} = (a + b\sqrt{5})^5$. Q.E.D.

LEMMA 9.7 *If c and d are odd, $\gcd(c,d) = 1$, $5 \mid d$, and $N(\frac{1}{2}c + \frac{1}{2}d\sqrt{5})$ is a fifth power, then there exist odd a and b such that $\frac{1}{2}c + \frac{1}{2}d\sqrt{5} = (\frac{1}{2}a + \frac{1}{2}b\sqrt{5})^5$.*
PROOF: If $c_1 = \frac{1}{2}(c - d)$ and $d_1 = d$, then $5 \mid d_1$, $\gcd(c_1, d_1) = 1$, and $\frac{1}{2}c + \frac{1}{2}d\sqrt{5} = \frac{1}{2}c + \frac{1}{2}d(2\omega - 1) = \frac{1}{2}(c - d) + d\omega = c_1 + d_1\omega$. Therefore, by Lemma 9.5, there exist a_1 and b_1 such that $\frac{1}{2}c + \frac{1}{2}d\sqrt{5} = c_1 + d_1\omega = (a_1 + b_1\omega)^5$. Modulo 2 we have $c_1 + d_1\omega = (a_1 + b_1\omega)^5 \equiv (a_1 + b_1\omega)^2 \equiv a_1^2 + b_1^2\omega^2 = a_1^2 + b_1^2 + b_1^2\omega$ (MOD

2). Thus, since d_1 is odd, then b_1 is odd; and so $a_1 + b_1\omega = \frac{1}{2}(2a_1 + b_1) + \frac{1}{2}b_1\sqrt{5} = \frac{1}{2}a + \frac{1}{2}b\sqrt{5}$, where a and b are odd. We therefore have $\frac{1}{2}c + \frac{1}{2}d\sqrt{5} = (\frac{1}{2}a + \frac{1}{2}b\sqrt{5})^5$, where a and b are odd. Q.E.D.

LEMMA 9.8 *There do not exist positive a and b with the following properties*:

(i) $\gcd(a, b) = 1$.

(ii) $10 \,|\, b$.

(iii) $5^4 \cdot 2b$ *and* $a^4 + 10a^2b^2 + 5b^4$ *are fifth powers.*

PROOF: The proof is by infinite descent. Specifically, we shall assume the existence of positive a and b having properties (i)–(iii) and show that this assumption would lead to positive a_1 and b_1 satisfying (i)–(iii) with $b > b_1$. Since this process could be repeated indefinitely, we would obtain an infinite sequence $b_1, b_2, \ldots$, of decreasing positive rational integers. As this is not possible, then the original assumption must be false, and the theorem would be proved.

Completing the square in the expression $a^4 + 10a^2b^2 + 5b^4$, we obtain $N(c + d\sqrt{5}) = a^4 + 10a^2b^2 + 5b^4$, where $c = a^2 + 5b^2$ and $d = 2b^2$. It is clear that $5 \,|\, d$, c is odd, d is even, and $\gcd(c, d) = 1$. Thus, by Lemma 9.6, we have $c + d\sqrt{5} = (a_1 + b_1\sqrt{5})^5$. Expanding $(a_1 + b_1\sqrt{5})^5$, we obtain

$$c = a_1^5 + 50a_1^3b_1^2 + 125a_1b_1^4,$$
$$d = 5a_1^4b_1 + 50a_1^2b_1^3 + 25b_1^5.$$

Numerous conclusions regarding a_1 and b_1 follow from the last two equations. Clearly, a_1 and b_1 are positive, since c and d are positive. The integers a_1 and b_1 are relatively prime, since c and d are relatively prime. Since c is odd, then a_1 is odd and b_1 is even. The integer a_1 is relatively prime to 5, for otherwise c and d would have the common factor 5. Since $d = 2b^2$, then $25 \,|\, d$, and so $5 \,|\, a_1^4b_1$. Consequently, as a_1 is relatively prime to 5, then $5 \,|\, b_1$. Thus we deduce that $a_1 > 0$, $b_1 > 0$, $\gcd(a_1, b_1) = 1$, and $10 \,|\, b_1$. Accordingly, the positive integers a_1 and b_1 satisfy Properties (i) and (ii) of the lemma. They also satisfy (iii), as we shall now show.

Since $d = 2b^2$ and $5^4 \cdot 2b$ is a fifth power, then $(5^4 \cdot 2b)^2 = 5^8 \cdot 2(2b^2) = 5^8 \cdot 2d$ is a fifth power. Thus $5^3 \cdot 2d$ is a fifth power. Using the expression for d in terms of a_1 and b_1, we conclude that $(5^4 \cdot 2b_1)(a_1^4 + 10a_1^2b_1^2 + 5b_1^4)$ is a fifth power. Since a_1 and b_1 satisfy Properties (i) and (ii), it is clear that the two factors in parentheses of the last expression are relatively prime; and so we conclude that $5^4 \cdot 2b_1$ and $a_1^4 + 10a_1^2b_1^2 + 5b_1^4$ are fifth powers. Hence the positive integers a_1 and b_1 satisfy (i)–(iii) of the lemma. All that remains is to show that $b > b_1$. This follows from $2b^2 = d = 5a_1^4b_1 + 50a_1^2b_1^3 + 25b_1^5 > 2b_1^2$, and the proof is complete. Q.E.D.

LEMMA 9.9 *There do not exist odd positive a and b with the following properties*:
(i) $\gcd(a,b) = 1$.
(ii) $5 \mid b$.
(iii) 5^4b and $2^{-4}(a^4 + 10a^2b^2 + 5b^4)$ *are fifth powers.*

PROOF: As in Lemma 9.8, we base the proof on infinite descent. Thus we assume that the odd positive integers a and b satisfy (i)–(iii). Completing the square in the expression $2^{-4}(a^4 + 10a^2b^2 + 5b^4)$, we obtain $N(\frac{1}{2}c + \frac{1}{2}d\sqrt{5}) = 2^{-4}(a^4 + 10a^2b^2 + 5b^4)$, where $c = \frac{1}{2}(a^2 + 5b^2)$ and $d = b^2$. Since $a^2 + 5b^2 \equiv 1 + 1 \cdot 1 = 2$ (mod 4), then c is odd. Also, it is clear that $5 \mid d$, c and d are odd, and $\gcd(c,d) = 1$. Thus, by Lemma 9.7, there exist odd a_1 and b_1 such that $\frac{1}{2}c + \frac{1}{2}d\sqrt{5} = (\frac{1}{2}a_1 + \frac{1}{2}b_1\sqrt{5})^5$. Expanding the expression $(\frac{1}{2}a_1 + \frac{1}{2}b_1\sqrt{5})^5$, we obtain

$$c = 2^{-4}(a_1^5 + 50a_1^3b_1^2 + 125a_1b_1^4),$$
$$d = 2^{-4}(5a_1^4b_1 + 50a_1^2b_1^3 + 25b_1^5).$$

Similar to the proof of Lemma 9.8, we deduce that $a_1 > 0$, $b_1 > 0$, $\gcd(a_1,b_1) = 1$, and $5 \mid b_1$. Accordingly, the odd positive integers a_1 and b_1 satisfy Properties (i) and (ii) of the lemma. They also satisfy (iii), as we shall now show.

Since $d = b^2$ and 5^4b is a fifth power, then $(5^4b)^2 = 5^8b^2 = 5^8d$ is a fifth power. Thus 5^3d is a fifth power. Using the expression for d in terms of a_1 and b_1, we conclude that $2^{-4} \cdot 5^4b_1(a_1^4 + 10a_1^2b_1^2 + 5b_1^4)$ is a fifth power. Since the odd positive integers a_1 and b_1 satisfy (i) and (ii), it is clear that 5^4b_1 and $2^{-4}(a_1^4 + 10a_1^2b_1^2 + 5b_1^4)$ are relatively prime; and so we conclude that 5^4b_1 and $2^{-4}(a_1^4 + 10a_1^2b_1^2 + 5b_1^4)$ are fifth powers. Hence the odd positive integers a_1 and b_1 satisfy (i)–(iii) of the lemma. All that remains is to show that $b > b_1$. This follows from

$$b^2 = d = 2^{-4}(5a_1^4 b_1 + 50a_1^2 b_1^3 + 25b_1^5) > b_1^2,$$

and the proof is complete. Q.E.D.

LEMMA 9.10 *There do not exist nonzero x, y, z such that* $\gcd(x, y, z) = 1$, $10 \mid z$, *and* $x^5 + y^5 + z^5 = 0$.

PROOF: Suppose, to the contrary, that such nonzero rational integers exist. Then $x^5 + y^5 = (-z)^5 = w^5$, so that $x^5 + y^5$ is a fifth power of a nonzero w which is divisible by 10. By Lemma 9.4, x and y are odd, since z is even. Therefore $x + y$ and $x - y$ are even, and so we have $x + y = 2r$ and $x - y = 2s$. Then $x = r + s$, $y = r - s$, and

$$w^5 = x^5 + y^5 = (r + s)^5 + (r - s)^5 \qquad (9.7)$$
$$= 2r(r^4 + 10r^2 s^2 + 5s^4).$$

The integers r and s are of opposite parity, because both x and y are odd; and $\gcd(r, s) = 1$, because any prime factor common to r and s would be a factor of x, y, and z. Also, $5 \mid r$, for if $5 \nmid r$, then $5 \nmid (r^4 + 10r^2 s^2 + 5s^4)$, and thus $5 \nmid w$, contrary to hypothesis. In summary, the integers r and s are relatively prime, of opposite parity, and $5 \mid r$. We thus have $r = 5u$, and so it follows from (9.7) that

$$(5^2 \cdot 2u)(s^4 + 50u^2 s^2 + 125u^4) = w^5. \qquad (9.8)$$

From the properties of r and s, it is easy to see that the two factors in parentheses on the left side of (9.8) are relatively prime. Hence we conclude that $5^2 \cdot 2u$ and $s^4 + 50u^2 s^2 + 125u^4$ are both fifth powers. Completing the square, we obtain $N(c + d\sqrt{5}) = s^4 + 50u^2 s^2 + 125u^4$, where $c = s^2 + 25u^2$ and $d = 10u^2$. Note that since $u \neq 0$, by (9.8), then c and d are both positive. Moreover, $N(c + d\sqrt{5})$ is a fifth power, $10 \mid d$, and since $r = 5u$ and s are relatively prime and of opposite parity, then $\gcd(c, d) = 1$. Consequently, by Lemma 9.6, we have $c + d\sqrt{5} = (a + b\sqrt{5})^5$. Expanding $(a + b\sqrt{5})^5$, we obtain

$$c = a^5 + 50a^3b^2 + 125ab^4,$$
$$d = 5a^4b + 50a^2b^3 + 25b^5.$$

As in the proof of Lemma 9.8, we deduce that $a > 0$, $b > 0$, $\gcd(a,b) = 1$, b is even, a is odd, and a is relatively prime to 5.

Since $5^2 \cdot 2u$ is a fifth power and $d = 10u^2$, then $(5^2 \cdot 2u)^2 = 5^3 \cdot 2(10u^2) = 5^3 \cdot 2d$ is a fifth power. Using the expression for d in terms of a and b, we conclude that $(5^4 \cdot 2b)(a^4 + 10a^2b^2 + 5b^4)$ is a fifth power. From the properties of a and b, we deduce that the two factors in parentheses of the last expression are relatively prime. Hence they are both fifth powers. Since $5^4 \cdot 2b$ is a fifth power, then $5 \mid b$. Thus the positive integers a and b satisfy (i)–(iii) of Lemma 9.8. Since this is impossible, it must be the case that the original assumption concerning x, y, z is false, and the lemma is established. Q.E.D.

LEMMA 9.11 *There do not exist nonzero x, y, z such that $\gcd(x, y, z) = 1$, $5 \mid z$, $2 \mid y$, and $x^5 + y^5 + z^5 = 0$.*

PROOF: Suppose, to the contrary, that such nonzero integers exist. Then $x^5 + y^5 = (-z)^5 = w^5$, so that $x^5 + y^5$ is a fifth power of a nonzero w which is divisible by 5. We set $r = x + y$ and $s = x - y$, so that $x = \frac{1}{2}(r + s)$ and $y = \frac{1}{2}(r - s)$. By Lemma 9.4, x is odd, and so r and s are odd. We have

$$w^5 = x^5 + y^5 = 2^{-5}\left[(r + s)^5 + (r - s)^5\right] \qquad (9.9)$$
$$= 2^{-4}r(r^4 + 10r^2s^2 + 5s^4).$$

Now $\gcd(r,s) = 1$, because any prime factor common to r and s would be a factor of x, y, and z. Also, $5 \mid r$, for if $5 \nmid r$, then $5 \nmid (r^4 + 10r^2s^2 + 5s^4)$, and thus $5 \nmid w^5$, contrary to hypothesis. In summary, the integers r and s are odd, $\gcd(r,s) = 1$, and $5 \mid r$. We thus have $r = 5u$, and so it follows from (9.9) that

$$2^{-4}(5^2u)(s^4 + 50u^2s^2 + 125u^4) = w^5. \qquad (9.10)$$

From the properties of r and s, it is clear that the two factors in parentheses on the left side of (9.10) are relatively prime. Hence we conclude that these two factors are both fifth powers. Completing the square, we obtain $N(\frac{1}{2}c + \frac{1}{2}d\sqrt{5}) = 2^{-4}(s^4 + 50u^2s^2 + 125u^4)$, where $c = \frac{1}{2}(s^2 + 25u^2)$ and $d = 5u^2$. Note that since $u \neq 0$, by (9.10), then c and d are positive. We have $s^2 + 25u^2 \equiv 1 + 1 \cdot 1 = 2$

(mod 4), so that c is odd. Since $d = 5u^2$, then d is odd and $5 \mid d$. Also, $\gcd(c, d) = 1$, since $\gcd(r, s) = 1$. In summary, c and d are odd positive integers, $\gcd(c, d) = 1$, $5 \mid d$, and $N(\frac{1}{2}c + \frac{1}{2}d\sqrt{5})$ is a fifth power. Thus, by Lemma 9.7, there exist odd a and b such that $\frac{1}{2}c + \frac{1}{2}d\sqrt{5} = (\frac{1}{2}a + \frac{1}{2}b\sqrt{5})^5$. Expanding $(\frac{1}{2}a + \frac{1}{2}b\sqrt{5})^5$, we obtain

$$c = 2^{-4}(a^5 + 50a^3b^2 + 125ab^4),$$

$$d = 2^{-4}(5a^4b + 50a^2b^3 + 25b^5).$$

The now familiar arguments show that a and b are odd positive integers, $\gcd(a, b) = 1$, and a is relatively prime to 5.

Since 5^2u is a fifth power and $d = 5u^2$, then $(5^2u)^2 = 5^4u^2 = 5^3(5u^2) = 5^3d$ is a fifth power. Using the expression for d in terms of a and b, we conclude that $2^{-4}(5^4b)(a^4 + 10a^2b^2 + 5b^4)$ is a fifth power. From the properties of a and b, we deduce that the two factors in parentheses of the last expression are relatively prime. Hence these two factors are both fifth powers. Since 5^4b is a fifth power, then $5 \mid b$. Thus the odd positive integers a and b satisfy (i)–(iii) of Lemma 9.9. Since this is impossible, it must be the case that the original assumption concerning x, y, z is false, and the lemma is established. Q.E.D.

THEOREM 9.12 *There do not exist nonzero x, y, z such that $x^5 + y^5 + z^5 = 0$.*

PROOF: It suffices to prove that there do not exist nonzero x, y, z such that $\gcd(x, y, z) = 1$ and $x^5 + y^5 + z^5 = 0$. We suppose the contrary. Then, by Lemma 9.4, exactly one integer in the set $S = \{x, y, z\}$ is divisible by 2, and exactly one integer in S is divisible by 5. Thus either 5 divides the even integer in S, or else 5 divides an odd integer in S. Since the preceding statement contradicts the results given in Lemmas 9.10 and 9.11, we are done. Q.E.D.

Divisibility Properties of the Fibonacci Numbers

We bring the chapter to a close by using the arithmetic theory of $\hat{Z}$ to derive some of the well-known divisibility properties of the Fibonacci numbers. Throughout this section we denote the Fibonacci number F_n by $F(n)$ and call n the *index* of $F(n)$. Also, p and q denote rational primes; and m, n, and r denote positive rational integers. The starting point for the investigations is the Binet formula for $F(n)$.

THEOREM 9.13 (Binet formula) $F(n) = (\omega^n - \overline{\omega}^n)/\sqrt{5}$.
PROOF: It follows from Theorem 3.3 that $\omega^n - \overline{\omega}^n = F(n-1) + F(n)\omega - [F(n+1) - F(n)\omega] = [F(n-1) - F(n+1)] + 2F(n)\omega = -F(n) + 2F(n)\omega = (-1 + 2\omega)F(n) = \sqrt{5}\,F(n)$. Q.E.D.

Some of the most important arithmetical properties of the Fibonacci numbers are listed in Theorem 9.18. We base a proof of this theorem on the properties given in Theorems 9.14, 9.15, and 9.17.

NOTATION $V(n) = \omega^{2n} - (-1)^n$.

THEOREM 9.14 $V(n) = \sqrt{5}\,\omega^n F(n)$.
PROOF: By Theorem 9.13, we have $\sqrt{5}\,\omega^n F(n) = \omega^n(\omega^n - \overline{\omega}^n) = \omega^{2n} - (\omega\overline{\omega})^n = \omega^{2n} - (-1)^n = V(n)$. Q.E.D.

THEOREM 9.15 *If* $m \mid n$, *then* $V(m) \mid V(n)$.
PROOF: Let $\alpha = \omega^{2m}$, $\beta = (-1)^m$, and $n = mt$, so that $V(m) = \alpha - \beta$ and $V(n) = \alpha^t - \beta^t$. Then $\gamma = \alpha^{t-1} + \alpha^{t-2}\beta + \ldots + \beta^{t-1}$ is an integer and $V(n) = \alpha^t - \beta^t = (\alpha - \beta)\gamma = V(m) \cdot \gamma$. Thus $V(m) \mid V(n)$. Q.E.D.

LEMMA 9.16 *If* $\omega^{2n} \equiv (-1)^n$ (MOD μ), *then* $\omega^{2na} \equiv (-1)^{na}$ (MOD μ) *for any rational integer a.*

PROOF: If $a \geq 0$, the result is immediate. If $a < 0$, then $\omega^{-2na} \equiv (-1)^{-na}$ (MOD μ). Multiplying both sides of the last congruence by the integer $(-1)^{na}\omega^{2na}$, we obtain $\omega^{2na} \equiv (-1)^{na}$ (MOD μ). Q.E.D.

THEOREM 9.17 *If $d = \gcd(n, m)$, then*

$$\mathrm{GCD}(V(n), V(m)) = V(d).$$

PROOF: Let $\delta = \mathrm{GCD}(V(n), V(m))$. Since $d = \gcd(n, m)$, there exist a and b such that $d = ma + nb$. Now $V(m) \equiv 0$ (MOD δ), so that $\omega^{2m} \equiv (-1)^m$ (MOD δ). Similarly, $\omega^{2n} \equiv (-1)^n$ (MOD δ). Thus, by Lemma 9.16, $\omega^{2ma} \equiv (-1)^{ma}$ (MOD δ) and $\omega^{2nb} \equiv (-1)^{nb}$ (MOD δ). Accordingly, $\omega^{2ma+2nb} \equiv (-1)^{ma+nb}$ (MOD δ), and since $d = ma + nb$, then $\omega^{2d} \equiv (-1)^d$ (MOD δ). Consequently, $V(d) = \omega^{2d} - (-1)^d \equiv 0$ (MOD δ); that is, $\delta \mid V(d)$. Conversely, since $d \mid n$ and $d \mid m$, then $V(d) \mid V(n)$ and $V(d) \mid V(m)$, by Theorem 9.15; and so $V(d) \mid \delta$. We thus conclude that $\delta = V(d)$ (up to associates). Q.E.D.

THEOREM 9.18
(i) *If $m \mid n$, then $F(m) \mid F(n)$.*
(ii) *If $d = \gcd(m, n)$, then $\gcd(F(m), F(n)) = F(d)$. In particular, if $\gcd(m, n) = 1$, then $\gcd(F(m), F(n)) = 1$.*
(iii) *If $\gcd(n, m) = 1$, then $F(m) \cdot F(n) \mid F(mn)$.*
(iv) *If $m > 2$, then $m \mid n$ if and only if $F(m) \mid F(n)$.*

PROOF: If $m \mid n$, then $V(m) \mid V(n)$ by Theorem 9.15. Thus, by Theorem 9.14, $\sqrt{5}\,\omega^m F(m)$ divides $\sqrt{5}\,\omega^n F(n)$; and since ω is a unit, then $F(m) \mid F(n)$. This establishes (i). By Theorem 9.14 and Theorem 9.17, we have that $\sqrt{5}\,\omega^d F(d) = V(d) = \mathrm{GCD}(V(m), V(n)) = \mathrm{GCD}(\sqrt{5}\,\omega^m F(m), \sqrt{5}\,\omega^n F(n)) = \sqrt{5}\,\cdot\mathrm{GCD}(F(m), F(n))$. Thus $F(d) = \mathrm{GCD}(F(m), F(n))$, and so $F(d) = \gcd(F(m), F(n))$. Consequently, (ii) is true. Assertion (iii) now follows from (i) and (ii), since $F(m) \mid F(mn)$, $F(n) \mid F(mn)$, and $\gcd(F(m), F(n)) = 1$. Half of (iv) follows from (i). Suppose now that $F(m) \mid F(n)$. Then, by (ii), we have $F(m) = \gcd(F(m), F(n)) = F(d)$, where $d = \gcd(m, n)$. Thus $F(m) = F(d)$; and if $m > 2$, we have $m = d = \gcd(m, n)$, so that $m \mid n$. Q.E.D.

COROLLARY 9.19 $\gcd(F(n), F(n + 1)) = 1$.
PROOF: We have $\gcd(n, n + 1) = 1$, and so, by Theorem 9.18(ii), $\gcd(F(n), F(n + 1)) = F(1) = 1$. Q.E.D.

If $m > 0$ is given, then we will see in Theorem 9.21 that $m \mid F(n)$ for some Fibonacci number $F(n)$. The remainder of this section deals with various aspects of this problem. In particular, we will be interested in the case $m = p^r$. The results obtained are well-known, but the theoretical development is rather novel. The key result we need from the arithmetic theory of $\hat{Z}$ is given by the following theorem.

THEOREM 9.20 *We have that* $m \mid F(n)$ *if and only if* ω^n *is congruent modulo* m *to a rational integer. Moreover, if* $m \mid F(n)$, *then* $\omega^n \equiv F(n - 1)$ (MOD m).
PROOF: If $m \mid F(n)$, then $\omega^n = F(n - 1) + F(n)\omega \equiv F(n - 1)$ (MOD m). Conversely, if $\omega^n \equiv a$ (MOD m), then $\bar{\omega}^n \equiv a$ (MOD m). We therefore have $\omega^n - \bar{\omega}^n \equiv 0$ (MOD m); and since $\omega^n - \bar{\omega}^n = \sqrt{5}\, F(n) = (-1 + 2\omega)F(n) = -F(n) + 2F(n)\omega$, it follows that $m \mid F(n)$. Q.E.D.

THEOREM 9.21 *If* $s = \text{ord}(\omega, m)$, *then* $m \mid F(s)$ *and* $m \mid F(\Phi(m))$. *Thus every* $m > 0$ *divides a Fibonacci number* $F(n)$, *where* $1 \le n \le m^2$.
PROOF: We have $\omega^s \equiv 1$ (MOD m) and, by the Euler–Fermat theorem $\omega^{\Phi(m)} \equiv 1$ (MOD m). Thus, by Theorem 9.20, $m \mid F(s)$ and $m \mid F(\Phi(m))$. The second assertion of the theorem follows from the observation that $1 \le \Phi(m) \le |N(m)| = m^2$. Q.E.D.

Vorobyov (Vorobyov [21], pp. 23–24) establishes the second statement of Theorem 9.21, but does not give a specific value n. His verification is the standard elementary proof of this result. A little later in Theorems 9.26 and 9.27 we will see that, except for $m = 1$ and $m = 2$, $\Phi(m)$ is not the smallest index n such that $m \mid F(n)$; and, in general, neither is $s = \text{ord}(\omega, m)$. For example, $\text{ord}(\omega, 7) = 16$, $\Phi(7) = 48$, and $7 \mid F(8)$. Brosseau (Brosseau [2]) gives $\text{ord}(\omega, p)$ for $p \le 269$. He denotes $\text{ord}(\omega, p)$ by $k(p)$ and calls $k(p)$ the *period* of p.

THEOREM 9.22 (Law of apparition) *If $p \equiv \pm 2$ (mod 5) and $q \equiv \pm 1$ (mod 5), then*
(i) $p \mid F(p + 1)$,
(ii) $q \mid F(q - 1)$,
(iii) $5 \mid F(5)$.
PROOF: By Theorem 5.51(i), $\omega^{p+1} \equiv N(\omega) = -1$ (MOD p). Thus, by Theorem 9.20, $p \mid F(p + 1)$. By Theorem 5.51(ii), $\omega^{q-1} \equiv 1$ (MOD q), and so $q \mid F(q - 1)$, by Theorem 9.20. Assertion (iii) is immediate, since $F(5) = 5$. Q.E.D.

THEOREM 9.23 *If $p^r \mid F(n)$, the $p^{r+1} \mid F(np)$.*
PROOF: Since $p^r \mid F(n)$, then $\omega^n \equiv a$ (MOD p^r), by Theorem 9.20. Thus $\omega^n = a + p^r \alpha$, and so $\omega^{np} = (a + p^r \alpha)^p \equiv a^p + pa^{p-1}p^r \alpha \equiv a^p$ (MOD p^{r+1}). It therefore follows from Theorem 9.20 that $p^{r+1} \mid F(np)$. Q.E.D.

THEOREM 9.24 *If $p \mid F(n)$, then $p^r \mid F(p^{r-1}n)$.*
PROOF: The proof is by induction on r. By hypothesis, the result holds for $r = 1$; and if $p^r \mid F(p^{r-1}n)$, then $p^{r+1} \mid F(p^r n)$, by Theorem 9.23. Q.E.D.

THEOREM 9.25 *If $p \equiv \pm 2$ (mod 5) and $q \equiv \pm 1$ (mod 5), then*
(i) $p^r \mid F(p^{r-1}(p + 1))$,
(ii) $q^r \mid F(q^{r-1}(q - 1))$,
(iii) $5^r \mid F(5^r)$.
PROOF: Immediate from Theorems 9.22 and 9.24. Q.E.D.

NOTATION If $p \equiv \pm 2$ (mod 5) and $q \equiv \pm 1$ (mod 5), then $T(1) = 1$, $T(p^r) = p^{r-1}(p + 1)$, $T(q^r) = q^{r-1}(q - 1)$, $T(5^r) = 5^r$; and if m has the rational prime decomposition $m = p_1^{c_1}p_2^{c_2} \cdots p_s^{c_s}$, then

$$T(m) = \text{lcm}(T(p_1^{c_1}), T(p_2^{c_2}), \ldots, T(p_s^{c_s})).$$

THEOREM 9.26 *We have $m \mid F(T(m))$.*
PROOF: The result is certainly true if $m = 1$. If $m > 1$, then let m have the rational prime decomposition $m = p_1^{c_1}p_2^{c_2} \cdots p_s^{c_s}$. Since $T(p_i^{c_i})$ divides $T(m)$, then $F(T(p_i^{c_i}))$ divides $F(T(m))$, by Theorem 9.18(i). Also, $p_i^{c_i}$ divides $F(T(p_i^{c_i}))$, by Theorem 9.25. Thus $p_i^{c_i}$

divides $F(T(m))$, and since the $p_i^{c_i}$ are pairwise relatively prime, then m divides $F(T(m))$. Q.E.D.

THEOREM 9.27 $T(m) \le \Phi(m)$ *with equality if and only if* $m = 1$ *or* $m = 2$.
PROOF: It is clear that $T(1) = \Phi(1) = 1$ and $T(2) = \Phi(2) = 3$. Now suppose that $m > 2$ and m has the rational prime decomposition $m = p_1^{c_1} p_2^{c_2} \cdots p_s^{c_s}$. We have $\Phi(m) = \prod \Phi(p_i^{c_i})$ and $T(m) \le \prod T(p_i^{c_i})$, where the products extend from $i = 1$ to $i = s$. Thus it suffices to show that $\prod T(p_i^{c_i}) < \prod \Phi(p_i^{c_i})$; and since $m > 2$, it is enough to show that $T(p_i^{c_i}) \le \Phi(p_i^{c_i})$ with inequality when $p_i^{c_i} \ne 2$. Since this is immediate from Lemma 5.43, we are done. Q.E.D.

DEFINITION   The **rank of apparition** of $m > 0$ is the smallest $t > 0$ such that $m \mid F(t)$. We also say that the index t is the **point of entry** of m in the Fibonacci numbers.

We note that Theorem 9.21 guarantees the existence of the rank of apparition of m. Brosseau (Brosseau [2]) gives the rank of apparition for each rational prime $p \le 269$; Jarden (Jarden [12]) gives the rank of apparition for each rational prime $p \le 1511$.

THEOREM 9.28   *The rank of apparition of* m *is the smallest* $t > 0$ *such that* ω^t *is congruent modulo* m *to a rational integer.*
PROOF: Immediate from Theorem 9.20. Q.E.D.

LEMMA 9.29   *The integer* $c + d\omega$ *is congruent modulo* m *to a rational integer if and only if* $m \mid d$.
PROOF: If $m \mid d$, then $c + d\omega \equiv c$ (MOD m). Conversely, if $c + d\omega \equiv a$ (MOD m), then $m \mid (c - a)$ and $m \mid d$. Q.E.D.

THEOREM 9.30   *If* t *is the rank of apparition of* m, *then* $m \mid F(n)$ *if and only if* $t \mid n$.
PROOF: First suppose that $t \mid n$. Then, by Theorem 9.18(i), we have $F(t) \mid F(n)$; and since $m \mid F(t)$, then $m \mid F(n)$.

Conversely, suppose that $m \mid F(n)$, so that $\omega^n \equiv b$ (MOD m), by Theorem 9.20. Since $n \ge t$, then $n = st + x$, where $s > 0$ and $0 \le x < t$. Thus, as $\omega^t \equiv a$ (MOD m) for some a with $\gcd(a, m) = 1$, we have $b \equiv \omega^n \equiv \omega^{st+x} \equiv a^s\omega^x$ (MOD m). Suppose $x \neq 0$. Then $\omega^x = c + d\omega$ and $m \nmid d$. (For if $m \mid d$, we would have $\omega^x \equiv c$ (MOD m), and so $m \mid F(x)$, a contradiction to the minimality of t.) Thus $b \equiv ca^s + da^s\omega$ (MOD m). This is impossible, by Lemma 9.29. (Since $\gcd(a,m) = 1$ and $m \nmid d$, then $m \nmid da^s$.) Accordingly, $x = 0$ and therefore $t \mid n$. Q.E.D.

NOTATION By $p^r \| n$, we mean that $p^r \mid n$ and $p^{r+1} \nmid n$.

LEMMA 9.31 *If $p^r \| F(n)$, then $p^{r+1} \mid F(nm)$ if and only if $p \mid m$.*
PROOF: Suppose $p \mid m$. We have that $p^{r+1} \mid F(np)$, by Theorem 9.23; and since $np \mid nm$, it follows from Theorem 9.18(i) that $F(np)$ divides $F(nm)$. Consequently, $p^{r+1} \mid F(nm)$.

Now suppose that $p^{r+1} \mid F(nm)$. Set $a = F(n-1)$ and $bp^r = F(n)$. Since $p^r \| F(n)$, then $\gcd(b, p) = 1$; and $\gcd(a, p) = 1$, since $F(n-1)$ and $F(n)$ are relatively prime. Therefore $\gcd(ab, p) = 1$. Also, $\omega^{nm} = (a + bp^r\omega)^m \equiv a^m + ma^{m-1}bp^r\omega$ (MOD p^{r+1}). Now $p^{r+1} \mid F(nm)$, and so, by Theorem 9.20, we have $a^m + ma^{m-1}bp^r\omega \equiv c$ (MOD p^{r+1}). But this means, by Lemma 9.29, that $p \mid ma^{m-1}b$; and since $\gcd(ab, p) = 1$, then $p \mid m$. Q.E.D.

LEMMA 9.32 *If $p^r \| F(n)$ and $\gcd(m, p) = 1$, then $p^{r+1} \mid F(nmp)$, and if $p^r \neq 2$, $p^{r+1} \| F(nmp)$.*
PROOF: Since $n \mid nm$, then $F(n) \mid F(nm)$, and so $p^r \mid F(nm)$. Thus $p^{r+1} \mid F(nmp)$, by Theorem 9.23. Also, since $\gcd(m, p) = 1$, then $p \nmid m$, so that $p^{r+1} \nmid F(nm)$, by Lemma 9.31. Accordingly, $p^r \| F(nm)$. Let $x = nm$. Then we have $p^r \| F(x)$, and we are to show that if $p^r \neq 2$, then $p^{r+1} \| F(xp)$. Of course, we already know that $p^{r+1} \mid F(xp)$, and so it only remains to show that $p^{r+2} \nmid F(xp)$.

Suppose first that $p > 2$. Set $a = F(x-1)$ and $bp^r = F(x)$. As in the proof of Lemma 9.31, we have $\gcd(ab, p) = 1$. Also, $\omega^{px} = (a + bp^r\omega)^p = a^p + p^{r+1}a^{p-1}b\omega + \alpha p^{r+2}$. Thus $\omega^{px} \equiv a^p + p^{r+1}a^{p-1}b\omega$ (MOD p^{r+2}), and as $p^{r+2} \nmid p^{r+1}a^{p-1}b$, then ω^{px} is not congruent modulo p^{r+2} to a rational integer (Lemma 9.29). Therefore, by Theorem 9.20, $p^{r+2} \nmid F(xp)$.

The proof for the exceptional case $p = 2$, $r > 1$ is exactly the same. (The condition $r > 1$ is needed to obtain the term αp^{r+2}.) Q.E.D.

THEOREM 9.33 (Law of repetition) *If $p^r \| F(n)$ and $\gcd(m, p)$ $= 1$, then for any $k \geq 0$, $p^{r+k} \mid F(nmp^k)$, and if $p^r \neq 2$, $p^{r+k} \| F(nmp^k)$.*

PROOF: A straightforward induction on k using Theorem 9.23 and Lemma 9.32. Q.E.D.

Theorem 9.33 is important in the theory of Fibonacci numbers, since if the behavior of p at its point of entry in the Fibonacci numbers is known, then the behavior of p for all other Fibonacci numbers can be predicted. We make this comment precise in the next two corollaries.

COROLLARY 9.34 *If t is the rank of apparition of $p \neq 2$, $p^r \| F(t)$, and $k \geq 0$, then $p^{r+k} \| F(n)$ if and only if $n = tmp^k$, where $\gcd(m, p) = 1$.*

PROOF: If $n = tmp^k$ and $\gcd(m, p) = 1$, then $p^{r+k} \| F(n)$, by Theorem 9.33. Now suppose that $p^{r+k} \| F(n)$. By Theorem 9.30, n is of the form $n = tmp^i$, where $i \geq 0$ and $\gcd(m, p) = 1$. It then follows from Theorem 9.33 that $i = k$. Q.E.D.

COROLLARY 9.35
(i) $2 \| F(n)$ if and only if $n = 3m$, where $\gcd(m, 2) = 1$.
(ii) If $k \geq 0$, then $2^{3+k} \| F(n)$ if and only if $n = 6(2^k)m$, where $\gcd(m, 2) = 1$.

PROOF: If $n = 3m$, then $2 \mid F(n)$ because $2 \mid F(3)$ and $F(3) \mid F(n)$. In addition, if $\gcd(2, m) = 1$, then $2^2 \nmid F(3m)$, by Lemma 9.31 (since $2 \| F(3)$). Conversely, suppose that $2 \| F(n)$. By Theorem 9.30, n is of the form $n = 3(2^i)m$, where $i \geq 0$ and $\gcd(2, m) = 1$. It then follows from Lemma 9.31 that $i = 0$. Thus (i) is true.

We note that $2^3 \| F(6)$. Thus, if $n = 6(2^k)m$ and $\gcd(m, 2) = 1$, then $2^{3+k} \| F(n)$, by Theorem 9.33. Conversely, suppose that $2^{3+k} \| F(n)$. By Theorem 9.30, n is of the form $n = 3(2^i)m$, where $i \geq 0$ and $\gcd(m, 2) = 1$. But $i \neq 0$, because then we would have $2 \| F(n)$, by (i). Thus n is of the form $n = 6(2^j)m$, where $j \geq 0$ and $\gcd(m, 2) = 1$. Therefore, since $2^3 \| F(6)$, it follows from Theorem 9.33 that $j = k$. This establishes (ii). Q.E.D.

The last two corollaries mark an appropriate stopping point for the intended rudimentary treatment of the divisibility properties of the Fibonacci numbers. Further properties could be obtained. One, in particular, is worthy of mention. A rational prime p is said to be a *primitive prime divisor* of $F(n)$ if the rank of apparition of p is n. It can be shown that every Fibonacci number $F(n)$, $n \neq 1, 2, 6, 12$, has at least one primitive prime divisor. Leon Mattics has based an unpublished proof of this result on the arithmetic theory of $\hat{Z}$. The proof is difficult and uses some results from Birkhoff and Vandiver [1] which have not been developed in this book. A generalized result along these lines can be found in Carmichael [5], but its proof is even more difficult.

Appendixes

Appendix A: Fibonacci Numbers $F_1 - F_{40}$

$F_1 = 1$

$F_2 = 1$

$F_3 = 2$

$F_4 = 3$

$F_5 = 5$

$F_6 = 8$

$F_7 = 13$

$F_8 = 21$

$F_9 = 34$

$F_{10} = 55$

$F_{11} = 89$

$F_{12} = 144$

$F_{13} = 233$

$F_{14} = 377$

$F_{15} = 610$

$F_{16} = 987$

$F_{17} = 1597$

$F_{18} = 2584$

$F_{19} = 4181$

$F_{20} = 6765$

$F_{21} = 10946$

$F_{22} = 17711$

$F_{23} = 28657$

$F_{24} = 46368$

$F_{25} = 75025$

$F_{26} = 121393$

$F_{27} = 196418$

$F_{28} = 317811$

$F_{29} = 514229$

$F_{30} = 832040$

$F_{31} = 1346269$

$F_{32} = 2178309$

$F_{33} = 3524578$

$F_{34} = 5702887$

$F_{35} = 9227465$

$F_{36} = 14930352$

$F_{37} = 24157817$

$F_{38} = 39088169$

$F_{39} = 63245986$

$F_{40} = 102334155$

Appendix B: A List of Primes

The following table lists for each rational prime p, $p < 32771$, a prime π_p of $Z(\omega)$ such that π_p divides p. In the table, an integer $c + d\omega$ is represented by the ordered pair (c, d), and beside each rational prime p is listed the corresponding π_p. If the given prime π_p is not p or $2 + \omega$, then π_p and $\overline{\pi_p}$ are the two nonassociated prime divisors of p; otherwise, the prime π_p is the only nonassociated prime divisor of p. Also, if $\pi_p \neq p$, then $N(\pi_p) = p$. For example, beside the rational prime 31 we find $(5, 2)$. Thus, $5 + 2\omega$ and $5 + 2\overline{\omega}$ are the nonassociated prime divisors of the rational prime 31, and $N(5 + 2\omega) = 31$.

The table was generated by a program written in BASIC language and run on a TRS-80 microcomputer. Utilizing the ideas found in Example 3.2, it took the microcomputer about 150 minutes to find and factor the 3512 rational primes less than 32771.

2 (2,0)	3 (3,0)	5 (2,1)
7 (7,0)	11 (3,1)	13 (13,0)
17 (17,0)	19 (4,1)	23 (23,0)
29 (5,1)	31 (5,2)	37 (37,0)
41 (6,1)	43 (43,0)	47 (47,0)
53 (53,0)	59 (7,2)	61 (7,3)
67 (67,0)	71 (8,1)	73 (73,0)
79 (8,3)	83 (83,0)	89 (9,1)
97 (97,0)	101 (9,4)	103 (103,0)
107 (107,0)	109 (10,1)	113 (113,0)
127 (127,0)	131 (11,1)	137 (137,0)
139 (11,2)	149 (11,4)	151 (11,5)
157 (157,0)	163 (163,0)	167 (167,0)
173 (173,0)	179 (12,5)	181 (13,1)
191 (13,2)	193 (193,0)	197 (197,0)
199 (13,3)	211 (13,6)	223 (223,0)
227 (227,0)	229 (14,3)	233 (233,0)
239 (15,1)	241 (14,5)	251 (15,2)
257 (257,0)	263 (263,0)	269 (15,4)
271 (16,1)	277 (277,0)	281 (15,7)
283 (283,0)	293 (293,0)	307 (307,0)
311 (16,5)	313 (313,0)	317 (317,0)
331 (17,3)	337 (337,0)	347 (347,0)
349 (17,5)	353 (353,0)	359 (17,7)
367 (367,0)	373 (373,0)	379 (19,1)
383 (383,0)	389 (18,5)	397 (397,0)
401 (18,7)	409 (19,3)	419 (20,1)

421 (19,4)	431 (19,5)	433 (433,0)
439 (19,6)	443 (443,0)	449 (19,8)
457 (457,0)	461 (21,1)	463 (463,0)
467 (467,0)	479 (21,2)	487 (487,0)
491 (20,7)	499 (20,9)	503 (503,0)
509 (21,4)	521 (21,5)	523 (523,0)
541 (22,3)	547 (547,0)	557 (557,0)
563 (563,0)	569 (22,5)	571 (23,2)
577 (577,0)	587 (587,0)	593 (593,0)
599 (24,1)	601 (22,9)	607 (607,0)
613 (613,0)	617 (617,0)	619 (23,5)
631 (23,6)	641 (23,7)	643 (643,0)
647 (647,0)	653 (653,0)	659 (23,10)
661 (23,11)	673 (673,0)	677 (677,0)
683 (683,0)	691 (25,3)	701 (26,1)
709 (25,4)	719 (24,11)	727 (727,0)
733 (733,0)	739 (25,6)	743 (743,0)
751 (25,7)	757 (757,0)	761 (25,8)
769 (25,9)	773 (773,0)	787 (787,0)
797 (797,0)	809 (26,7)	811 (28,1)
821 (27,4)	823 (823,0)	827 (827,0)
829 (26,9)	839 (27,5)	853 (853,0)
857 (857,0)	859 (28,3)	863 (863,0)
877 (877,0)	881 (27,8)	883 (883,0)
887 (887,0)	907 (907,0)	911 (27,13)
919 (29,3)	929 (30,1)	937 (937,0)
941 (29,4)	947 (947,0)	953 (953,0)
967 (967,0)	971 (28,11)	977 (977,0)
983 (983,0)	991 (31,1)	997 (997,0)
1009 (29,8)	1013 (1013,0)	1019 (31,2)
1021 (29,9)	1031 (29,10)	1033 (1033,0)
1039 (29,11)	1049 (29,13)	1051 (29,14)
1061 (30,7)	1063 (1063,0)	1069 (31,4)
1087 (1087,0)	1091 (31,5)	1093 (1093,0)
1097 (1097,0)	1103 (1103,0)	1109 (30,11)
1117 (1117,0)	1123 (1123,0)	1129 (31,7)
1151 (33,2)	1153 (1153,0)	1163 (1163,0)
1171 (31,10)	1181 (31,11)	1187 (1187,0)
1193 (1193,0)	1201 (31,15)	1213 (1213,0)
1217 (1217,0)	1223 (1223,0)	1229 (33,5)
1231 (32,9)	1237 (1237,0)	1249 (34,3)
1259 (35,1)	1277 (1277,0)	1279 (32,15)
1283 (1283,0)	1289 (33,8)	1291 (35,2)
1297 (1297,0)	1301 (34,5)	1303 (1303,0)
1307 (1307,0)	1319 (33,10)	1321 (35,3)
1327 (1327,0)	1361 (33,16)	1367 (1367,0)
1373 (1373,0)	1381 (34,9)	1399 (35,6)
1409 (34,11)	1423 (1423,0)	1427 (1427,0)
1429 (34,13)	1433 (1433,0)	1439 (37,2)
1447 (1447,0)	1451 (36,5)	1453 (1453,0)
1459 (35,9)	1471 (37,3)	1481 (38,1)
1483 (1483,0)	1487 (1487,0)	1489 (35,11)
1493 (1493,0)	1499 (36,7)	1511 (35,13)
1523 (1523,0)	1531 (35,17)	1543 (1543,0)
1549 (38,3)	1553 (1553,0)	1559 (39,1)

1567 (1567,0)	1571 (36,11)	1579 (37,7)
1583 (1583,0)	1597 (1597,0)	1601 (37,8)
1607 (1607,0)	1609 (38,5)	1613 (1613,0)
1619 (36,17)	1621 (37,9)	1627 (1627,0)
1637 (1637,0)	1657 (1657,0)	1663 (1663,0)
1667 (1667,0)	1669 (37,12)	1693 (1693,0)
1697 (1697,0)	1699 (37,15)	1709 (37,17)
1721 (41,1)	1723 (1723,0)	1733 (1733,0)
1741 (38,11)	1747 (1747,0)	1753 (1753,0)
1759 (41,2)	1777 (1777,0)	1783 (1783,0)
1787 (1787,0)	1789 (38,15)	1801 (38,17)
1811 (39,10)	1823 (1823,0)	1831 (40,7)
1847 (1847,0)	1861 (41,5)	1867 (1867,0)
1871 (39,14)	1873 (1873,0)	1877 (1877,0)
1879 (40,9)	1889 (39,16)	1901 (39,19)
1907 (1907,0)	1913 (1913,0)	1931 (43,2)
1933 (1933,0)	1949 (42,5)	1951 (40,13)
1973 (1973,0)	1979 (44,1)	1987 (1987,0)
1993 (1993,0)	1997 (1997,0)	1999 (40,19)
2003 (2003,0)	2011 (41,11)	2017 (2017,0)
2027 (2027,0)	2029 (41,12)	2039 (43,5)
2053 (2053,0)	2063 (2063,0)	2069 (45,1)
2081 (41,16)	**2083 (2083,0)**	**2087 (2087,0)**
2089 (41,17)	**2099 (41,19)**	**2111 (45,2)**
2113 (2113,0)	**2129 (43,8)**	**2131 (44,5)**
2137 (2137,0)	**2141 (42,13)**	**2143 (2143,0)**
2153 (2153,0)	**2161 (46,1)**	**2179 (43,10)**
2203 (2203,0)	**2207 (2207,0)**	**2213 (2213,0)**
2221 (43,12)	**2237 (2237,0)**	**2239 (43,13)**
2243 (2243,0)	**2251 (44,9)**	**2267 (2267,0)**
2269 (43,15)	**2273 (2273,0)**	**2281 (43,16)**
2287 (2287,0)	**2293 (2293,0)**	**2297 (2297,0)**
2309 (43,20)	**2311 (43,21)**	**2333 (2333,0)**
2339 (44,13)	**2341 (47,3)**	**2347 (2347,0)**
2351 (48,1)	**2357 (2357,0)**	**2371 (44,15)**
2377 (2377,0)	**2381 (47,4)**	**2383 (2383,0)**
2389 (46,7)	**2393 (2393,0)**	**2399 (45,11)**
2411 (44,19)	**2417 (2417,0)**	**2423 (2423,0)**
2437 (2437,0)	**2441 (45,13)**	**2447 (2447,0)**
2459 (45,14)	**2467 (2467,0)**	**2473 (2473,0)**
2477 (2477,0)	**2503 (2503,0)**	**2521 (47,8)**
2531 (45,22)	**2539 (49,3)**	**2543 (2543,0)**
2549 (50,1)	**2551 (47,9)**	**2557 (2557,0)**
2579 (47,10)	**2591 (48,7)**	**2593 (2593,0)**
2609 (46,17)	**2617 (2617,0)**	**2621 (49,5)**
2633 (2633,0)	**2647 (2647,0)**	**2657 (2657,0)**
2659 (49,6)	**2663 (2663,0)**	**2671 (47,14)**
2677 (2677,0)	**2683 (2683,0)**	**2687 (2687,0)**
2689 (47,15)	**2693 (2693,0)**	**2699 (51,2)**
2707 (2707,0)	**2711 (48,11)**	**2713 (2713,0)**
2719 (47,17)	**2729 (49,8)**	**2731 (47,18)**
2741 (47,19)	**2749 (47,20)**	**2753 (2753,0)**
2767 (2767,0)	**2777 (2777,0)**	**2789 (51,4)**
2791 (49,10)	**2797 (2797,0)**	**2801 (50,7)**
2803 (2803,0)	**2819 (49,11)**	**2833 (2833,0)**

```
2837 (2837,0)    2843 (2843,0)    2851 (52,3)
2857 (2857,0)    2861 (53,1)      2879 (48,23)
2887 (2887,0)    2897 (2897,0)    2903 (2903,0)
2909 (51,7)      2917 (2917,0)    2927 (2927,0)
2939 (52,5)      2953 (2953,0)    2957 (2957,0)
2963 (2963,0)    2969 (54,1)      2971 (49,19)
2999 (49,23)     3001 (49,24)     3011 (51,10)
3019 (52,7)      3023 (3023,0)    3037 (3037,0)
3041 (51,11)     3049 (53,5)      3061 (50,17)
3067 (3067,0)    3079 (55,1)      3083 (3083,0)
3089 (50,19)     3109 (50,21)     3119 (51,14)
3121 (50,23)     3137 (3137,0)    3163 (3163,0)
3167 (3167,0)    3169 (53,8)      3181 (55,3)
3187 (3187,0)    3191 (56,1)      3203 (3203,0)
3209 (51,19)     3217 (3217,0)    3221 (51,20)
3229 (55,4)      3251 (51,25)     3253 (3253,0)
3257 (3257,0)    3259 (52,15)     3271 (53,11)
3299 (52,17)     3301 (53,12)     3307 (3307,0)
3313 (3313,0)    3319 (55,6)      3323 (3323,0)
3329 (53,13)     3331 (52,19)     3343 (3343,0)
3347 (3347,0)    3359 (57,2)      3361 (55,7)
3371 (52,23)     3373 (3373,0)    3389 (54,11)
3391 (56,5)      3407 (3407,0)    3413 (3413,0)
3433 (3433,0)    3449 (54,13)     3457 (3457,0)
3461 (57,4)      3463 (3463,0)    3467 (3467,0)
3469 (53,20)     3491 (53,22)     3499 (53,23)
3511 (53,26)     3517 (3517,0)    3527 (3527,0)
3529 (58,3)      3533 (3533,0)    3539 (59,1)
3541 (55,12)     3547 (3547,0)    3557 (3557,0)
3559 (56,9)      3571 (55,13)     3581 (54,19)
3583 (3583,0)    3593 (3593,0)    3607 (3607,0)
3613 (3613,0)    3617 (3617,0)    3623 (3623,0)
3631 (56,11)     3637 (3637,0)    3643 (3643,0)
3659 (60,1)      3671 (55,17)     3673 (3673,0)
3677 (3677,0)    3691 (55,18)     3697 (3697,0)
3701 (59,4)      3709 (55,19)     3719 (57,10)
3727 (3727,0)    3733 (3733,0)    3739 (55,21)
3761 (55,23)     3767 (3767,0)    3769 (55,24)
3779 (55,26)     3793 (3793,0)    3797 (3797,0)
3803 (3803,0)    3821 (57,13)     3823 (3823,0)
3833 (3833,0)    3847 (3847,0)    3851 (57,14)
3853 (3853,0)    3863 (3863,0)    3877 (3877,0)
3881 (58,11)     3889 (59,8)      3907 (3907,0)
3911 (56,25)     3917 (3917,0)    3919 (56,27)
3923 (3923,0)    3929 (57,17)     3931 (59,9)
3943 (3943,0)    3947 (3947,0)    3967 (3967,0)
3989 (57,20)     4001 (61,5)      4003 (4003,0)
4007 (4007,0)    4013 (4013,0)    4019 (57,22)
4021 (62,3)      4027 (4027,0)    4049 (57,25)
4051 (61,6)      4057 (4057,0)    4073 (4073,0)
4079 (59,13)     4091 (63,2)      4093 (4093,0)
4099 (61,7)      4111 (59,14)     4127 (4127,0)
4129 (62,5)      4133 (4133,0)    4139 (60,11)
4153 (4153,0)    4157 (4157,0)    4159 (64,1)
4177 (4177,0)    4201 (58,27)     4211 (60,13)
```

4217 (4217,0)	4219 (59,18)	4229 (62,7)
4231 (61,10)	4241 (59,19)	4243 (4243,0)
4253 (4253,0)	4259 (63,5)	4261 (59,20)
4271 (61,11)	4273 (4273,0)	4283 (4283,0)
4289 (65,1)	4297 (4297,0)	4327 (4327,0)
4337 (4337,0)	4339 (59,26)	4349 (59,28)
4357 (4357,0)	4363 (4363,0)	4373 (4373,0)
4391 (64,5)	4397 (4397,0)	4409 (63,8)
4421 (66,1)	4423 (4423,0)	4441 (61,16)
4447 (4447,0)	4451 (60,23)	4457 (4457,0)
4463 (4463,0)	4481 (62,13)	4483 (4483,0)
4493 (4493,0)	4507 (4507,0)	4513 (4513,0)
4517 (4517,0)	4519 (61,19)	4523 (4523,0)
4547 (4547,0)	4549 (62,15)	4561 (61,21)
4567 (4567,0)	4583 (4583,0)	4591 (64,9)
4597 (4597,0)	4603 (4603,0)	4621 (61,25)
4637 (4637,0)	4639 (61,27)	4643 (4643,0)
4649 (61,29)	4651 (61,30)	4657 (4657,0)
4663 (4663,0)	4673 (4673,0)	4679 (64,11)
4691 (68,1)	4703 (4703,0)	4721 (63,16)
4723 (4723,0)	4729 (65,9)	4733 (4733,0)
4751 (63,17)	4759 (64,13)	4783 (4783,0)
4787 (4787,0)	4789 (62,27)	4793 (4793,0)
4799 (67,5)	4801 (62,29)	4813 (4813,0)
4817 (4817,0)	4831 (64,15)	4861 (65,12)
4871 (63,22)	4877 (4877,0)	4889 (63,23)
4903 (4903,0)	4909 (67,7)	4919 (63,25)
4931 (63,26)	4933 (4933,0)	4937 (4937,0)
4943 (4943,0)	4951 (64,19)	4957 (4957,0)
4967 (4967,0)	4969 (70,1)	4973 (4973,0)
4987 (4987,0)	4993 (4993,0)	4999 (64,21)
5003 (5003,0)	5009 (65,16)	5011 (67,9)
5021 (69,4)	5023 (5023,0)	5039 (64,23)
5051 (68,7)	5059 (67,10)	5077 (5077,0)
5081 (69,5)	5087 (5087,0)	5099 (65,19)
5101 (70,3)	5107 (5107,0)	5113 (5113,0)
5119 (64,31)	5147 (5147,0)	5153 (5153,0)
5167 (5167,0)	5171 (65,22)	5179 (71,2)
5189 (66,17)	5197 (5197,0)	5209 (65,24)
5227 (5227,0)	5231 (67,14)	5233 (5233,0)
5237 (5237,0)	5261 (65,28)	5273 (5273,0)
5279 (65,31)	5281 (65,32)	5297 (5297,0)
5303 (5303,0)	5309 (71,4)	5323 (5323,0)
5333 (5333,0)	5347 (5347,0)	5351 (69,10)
5381 (66,25)	5387 (5387,0)	5393 (5393,0)
5399 (69,11)	5407 (5407,0)	5413 (5413,0)
5417 (5417,0)	5419 (68,15)	5431 (71,6)
5437 (5437,0)	5441 (66,31)	5443 (5443,0)
5449 (70,9)	5471 (73,2)	5477 (5477,0)
5479 (67,22)	5483 (5483,0)	5501 (67,23)
5503 (5503,0)	5507 (5507,0)	5519 (72,5)
5521 (67,24)	5527 (5527,0)	5531 (69,14)
5557 (5557,0)	5563 (5563,0)	5569 (67,27)
5573 (5573,0)	5581 (67,28)	5591 (67,29)
5623 (5623,0)	5639 (72,7)	5641 (70,13)

5647 (5647,0)	5651 (71,10)	5653 (5653,0)
5657 (5657,0)	5659 (68,23)	5669 (73,5)
5683 (5683,0)	5689 (74,3)	5693 (5693,0)
5701 (71,11)	5711 (69,19)	5717 (5717,0)
5737 (5737,0)	5741 (69,20)	5743 (5743,0)
5749 (71,12)	5779 (68,33)	5783 (5783,0)
5791 (73,7)	5801 (70,17)	5807 (5807,0)
5813 (5813,0)	5821 (74,5)	5827 (5827,0)
5839 (71,14)	5843 (5843,0)	5849 (73,8)
5851 (76,1)	5857 (5857,0)	5861 (69,25)
5867 (5867,0)	5869 (70,19)	5879 (69,26)
5881 (71,15)	5897 (5897,0)	5903 (5903,0)
5923 (5923,0)	5927 (5927,0)	5939 (69,31)
5953 (5953,0)	5981 (70,23)	5987 (5987,0)
6007 (6007,0)	6011 (73,11)	6029 (71,19)
6037 (6037,0)	6043 (6043,0)	6047 (6047,0)
6053 (6053,0)	6067 (6067,0)	6073 (6073,0)
6079 (77,2)	6089 (70,29)	6091 (71,21)
6101 (75,7)	6113 (6113,0)	6121 (70,33)
6131 (76,5)	6133 (6133,0)	6143 (6143,0)
6151 (77,3)	6163 (6163,0)	6173 (6173,0)
6197 (6197,0)	6199 (73,15)	6203 (6203,0)
6211 (71,26)	6217 (6217,0)	6221 (77,4)
6229 (71,27)	6247 (6247,0)	6257 (6257,0)
6263 (6263,0)	6269 (74,13)	6271 (71,30)
6277 (6277,0)	6287 (6287,0)	6299 (71,34)
6301 (71,35)	6311 (72,23)	6317 (6317,0)
6323 (6323,0)	6329 (75,11)	6337 (6337,0)
6343 (6343,0)	6353 (6353,0)	6359 (72,25)
6361 (74,15)	6367 (6367,0)	6373 (6373,0)
6379 (76,9)	6389 (73,20)	6397 (6397,0)
6421 (73,21)	6427 (6427,0)	6449 (78,5)
6451 (73,22)	6469 (79,3)	6473 (6473,0)
6481 (77,8)	6491 (76,11)	6521 (74,19)
6529 (73,25)	6547 (6547,0)	6551 (73,26)
6553 (6553,0)	6563 (6563,0)	6569 (75,16)
6571 (73,27)	6577 (6577,0)	6581 (78,7)
6599 (77,10)	6607 (6607,0)	6619 (73,30)
6637 (6637,0)	6653 (6653,0)	6659 (73,35)
6661 (73,36)	6673 (6673,0)	6679 (79,6)
6689 (75,19)	6691 (76,15)	6701 (74,25)
6703 (6703,0)	6709 (77,12)	6719 (81,2)
6733 (6733,0)	6737 (6737,0)	6761 (77,13)
6763 (6763,0)	6779 (76,17)	6781 (74,29)
6791 (75,22)	6793 (6793,0)	6803 (6803,0)
6823 (6823,0)	6827 (6827,0)	6829 (74,33)
6833 (6833,0)	6841 (74,35)	6857 (6857,0)
6863 (6863,0)	6869 (81,4)	6871 (79,9)
6883 (6883,0)	6899 (75,26)	6907 (6907,0)
6911 (80,7)	6917 (6917,0)	6947 (6947,0)
6949 (77,17)	6959 (75,29)	6961 (82,3)
6967 (6967,0)	6971 (83,1)	6977 (6977,0)
6983 (6983,0)	6991 (77,18)	6997 (6997,0)
7001 (75,32)	7013 (7013,0)	7019 (75,34)
7027 (7027,0)	7039 (80,9)	7043 (7043,0)

7057 (7057,0)	7069 (77,20)	7079 (81,7)
7103 (7103,0)	7109 (82,5)	7121 (78,17)
7127 (7127,0)	7129 (83,3)	7151 (79,14)
7159 (80,11)	7177 (7177,0)	7187 (7187,0)
7193 (7193,0)	7207 (7207,0)	7211 (76,35)
7213 (7213,0)	7219 (76,37)	7229 (77,25)
7237 (7237,0)	7243 (7243,0)	7247 (7247,0)
7253 (7253,0)	7283 (7283,0)	7297 (7297,0)
7307 (7307,0)	7309 (85,1)	7321 (77,29)
7331 (81,11)	7333 (7333,0)	7349 (78,23)
7351 (83,6)	7369 (77,32)	7393 (7393,0)
7411 (77,38)	7417 (7417,0)	7433 (7433,0)
7451 (84,5)	7457 (7457,0)	7459 (79,21)
7477 (7477,0)	7481 (86,1)	7487 (7487,0)
7489 (83,8)	7499 (81,14)	7507 (7507,0)
7517 (7517,0)	7523 (7523,0)	7529 (79,23)
7537 (7537,0)	7541 (78,31)	7547 (7547,0)
7549 (85,4)	7559 (80,19)	7561 (79,24)
7573 (7573,0)	7577 (7577,0)	7583 (7583,0)
7589 (78,35)	7591 (79,25)	7603 (7603,0)
7607 (7607,0)	7621 (82,13)	7639 (80,21)
7643 (7643,0)	7649 (81,17)	7669 (79,28)
7673 (7673,0)	7681 (83,11)	7687 (7687,0)
7691 (79,29)	7699 (85,6)	7703 (7703,0)
7717 (7717,0)	7723 (7723,0)	7727 (7727,0)
7741 (83,12)	7753 (7753,0)	7757 (7757,0)
7759 (79,33)	7789 (79,36)	7793 (7793,0)
7817 (7817,0)	7823 (7823,0)	7829 (82,17)
7841 (85,8)	7853 (7853,0)	7867 (7867,0)
7873 (7873,0)	7877 (7877,0)	7879 (80,29)
7883 (7883,0)	7901 (87,4)	7907 (7907,0)
7919 (80,31)	7927 (7927,0)	7933 (7933,0)
7937 (7937,0)	7949 (86,7)	7951 (80,33)
7963 (7963,0)	7993 (7993,0)	8009 (89,1)
8011 (83,17)	8017 (8017,0)	8039 (85,11)
8053 (8053,0)	8059 (83,18)	8069 (81,29)
8081 (82,23)	8087 (8087,0)	8089 (86,9)
8093 (8093,0)	8101 (85,12)	8111 (81,31)
8117 (8117,0)	8123 (8123,0)	8147 (8147,0)
8161 (85,13)	8167 (8167,0)	8171 (81,35)
8179 (89,3)	8191 (83,21)	8209 (82,27)
8219 (85,14)	8221 (86,11)	8231 (83,22)
8233 (8233,0)	8237 (8237,0)	8243 (8243,0)
8263 (8263,0)	8269 (83,23)	8273 (8273,0)
8287 (8287,0)	8291 (84,19)	8293 (8293,0)
8297 (8297,0)	8311 (88,7)	8317 (8317,0)
8329 (85,16)	8353 (8353,0)	8363 (8363,0)
8369 (82,35)	8377 (8377,0)	8387 (8387,0)
8389 (82,37)	8419 (89,6)	8423 (8423,0)
8429 (83,28)	8431 (85,18)	8443 (8443,0)
8447 (8447,0)	8461 (86,15)	8467 (8467,0)
8501 (83,31)	8513 (8513,0)	8521 (83,32)
8527 (8527,0)	8537 (8537,0)	8539 (83,33)
8543 (8543,0)	8563 (8563,0)	8573 (8573,0)
8581 (83,36)	8597 (8597,0)	8599 (83,38)

8609 (83,40)	8623 (8623,0)	8627 (8627,0)
8629 (91,4)	8641 (89,9)	8647 (8647,0)
8663 (8663,0)	8669 (86,19)	8677 (8677,0)
8681 (90,7)	8689 (85,24)	8693 (8693,0)
8699 (84,31)	8707 (8707,0)	8713 (8713,0)
8719 (88,13)	8731 (92,3)	8737 (8737,0)
8741 (93,1)	8747 (8747,0)	8753 (8753,0)
8761 (86,21)	8779 (89,11)	8783 (8783,0)
8803 (8803,0)	8807 (8807,0)	8819 (84,41)
8821 (85,28)	8831 (93,2)	8837 (8837,0)
8839 (88,15)	8849 (85,29)	8861 (87,19)
8863 (8863,0)	8867 (8867,0)	8887 (8887,0)
8893 (8893,0)	8923 (8923,0)	8929 (94,1)
8933 (8933,0)	8941 (85,33)	8951 (88,17)
8963 (8963,0)	8969 (90,11)	8971 (89,14)
8999 (87,22)	9001 (85,37)	9007 (9007,0)
9011 (85,38)	9013 (9013,0)	9029 (85,41)
9041 (87,23)	9043 (9043,0)	9049 (86,29)
9059 (92,7)	9067 (9067,0)	9091 (91,10)
9103 (9103,0)	9109 (94,3)	9127 (9127,0)
9133 (9133,0)	9137 (9137,0)	9151 (88,21)
9157 (9157,0)	9161 (91,11)	9173 (9173,0)
9181 (86,35)	9187 (9187,0)	9199 (89,18)
9203 (9203,0)	9209 (86,37)	9221 (87,28)
9227 (9227,0)	9277 (9277,0)	9241 (86,41)
9257 (9257,0)	9277 (9277,0)	9281 (94,5)
9283 (9283,0)	9293 (9293,0)	9311 (96,1)
9319 (88,25)	9323 (9323,0)	9337 (9337,0)
9341 (90,17)	9343 (9343,0)	9349 (89,21)
9371 (87,34)	9377 (9377,0)	9391 (88,27)
9397 (9397,0)	9403 (9403,0)	9413 (9413,0)
9419 (87,37)	9421 (91,15)	9431 (87,38)
9433 (9433,0)	9437 (9437,0)	9439 (89,23)
9461 (87,43)	9463 (9463,0)	9467 (9467,0)
9473 (9473,0)	9479 (93,10)	9491 (92,13)
9497 (9497,0)	9511 (88,31)	9521 (89,25)
9533 (9533,0)	9539 (91,17)	9547 (9547,0)
9551 (93,11)	9587 (9587,0)	9601 (94,9)
9613 (9613,0)	9619 (92,15)	9623 (9623,0)
9629 (89,28)	9631 (88,37)	9643 (9643,0)
9649 (91,19)	9661 (89,29)	9677 (9677,0)
9679 (88,43)	9689 (93,13)	9697 (9697,0)
9719 (89,31)	9721 (95,8)	9733 (9733,0)
9739 (92,17)	9743 (9743,0)	9749 (94,11)
9767 (9767,0)	9769 (89,33)	9781 (97,4)
9787 (9787,0)	9791 (89,34)	9803 (9803,0)
9811 (89,35)	9817 (9817,0)	9829 (89,36)
9833 (9833,0)	9839 (96,7)	9851 (92,19)
9857 (9857,0)	9859 (89,38)	9871 (89,39)
9883 (9883,0)	9887 (9887,0)	9901 (89,44)
9907 (9907,0)	9923 (9923,0)	9929 (90,31)
9931 (91,25)	9941 (93,17)	9949 (95,11)
9967 (9967,0)	9973 (9973,0)	10007 (10007,0)
10009 (91,27)	10037 (10037,0)	10039 (97,7)
10061 (90,37)	10067 (10067,0)	10069 (98,5)

10079 (91,29)	10091 (95,13)	10093 (10093,0)
10099 (100,1)	10103 (10103,0)	10111 (91,30)
10133 (10133,0)	10139 (92,25)	10141 (91,31)
10151 (96,11)	10159 (95,14)	10163 (10163,0)
10169 (91,32)	10177 (10177,0)	10181 (99,4)
10193 (10193,0)	10211 (93,22)	10223 (10223,0)
10243 (10243,0)	10247 (10247,0)	10253 (10253,0)
10259 (93,23)	10267 (10267,0)	10271 (99,5)
10273 (10273,0)	10289 (95,16)	10301 (101,1)
10303 (10303,0)	10313 (10313,0)	10321 (91,40)
10331 (91,41)	10333 (10333,0)	10337 (10337,0)
10343 (10343,0)	10357 (10357,0)	10369 (94,21)
10391 (93,26)	10399 (101,2)	10427 (10427,0)
10429 (97,12)	10433 (10433,0)	10453 (10453,0)
10457 (10457,0)	10459 (92,35)	10463 (10463,0)
10477 (10477,0)	10487 (10487,0)	10499 (92,37)
10501 (97,13)	10513 (10513,0)	10529 (99,8)
10531 (92,39)	10559 (96,17)	10567 (10567,0)
10589 (101,4)	10597 (10597,0)	10601 (93,32)
10607 (10607,0)	10613 (10613,0)	10627 (10627,0)
10631 (95,22)	10639 (97,15)	10651 (100,7)
10657 (10657,0)	10663 (10663,0)	10667 (10667,0)
10687 (10687,0)	10691 (99,10)	10709 (98,13)
10711 (103,1)	10723 (10723,0)	10729 (95,24)
10733 (10733,0)	10739 (93,38)	10753 (10753,0)
10771 (101,6)	10781 (93,41)	10789 (94,31)
10799 (93,43)	10831 (97,18)	10837 (10837,0)
10847 (10847,0)	10853 (10853,0)	10859 (101,7)
10861 (95,27)	10867 (10867,0)	10883 (10883,0)
10889 (102,5)	10891 (97,19)	10903 (10903,0)
10909 (103,3)	10937 (10937,0)	10939 (95,29)
10949 (97,20)	10957 (10957,0)	10973 (10973,0)
10979 (100,11)	10987 (10987,0)	10993 (10993,0)
11003 (11003,0)	11027 (11027,0)	11047 (11047,0)
11057 (11057,0)	11059 (97,22)	11069 (102,7)
11071 (95,33)	11083 (11083,0)	11087 (11087,0)
11093 (11093,0)	11113 (11113,0)	11117 (11117,0)
11119 (104,3)	11131 (100,13)	11149 (95,36)
11159 (96,29)	11161 (97,24)	11171 (95,37)
11173 (11173,0)	11177 (11177,0)	11197 (11197,0)
11213 (11213,0)	11239 (95,41)	11243 (11243,0)
11251 (95,42)	11257 (11257,0)	11261 (95,43)
11273 (11273,0)	11279 (95,46)	11287 (11287,0)
11299 (97,27)	11311 (104,5)	11317 (11317,0)
11321 (99,19)	11329 (98,23)	11351 (96,35)
11353 (11353,0)	11369 (103,8)	11383 (11383,0)
11393 (11393,0)	11399 (96,37)	11411 (100,17)
11423 (11423,0)	11437 (11437,0)	11443 (11443,0)
11447 (11447,0)	11467 (11467,0)	11471 (96,41)
11483 (11483,0)	11489 (97,32)	11491 (101,15)
11497 (11497,0)	11503 (11503,0)	11519 (96,47)
11527 (11527,0)	11549 (99,23)	11551 (97,34)
11579 (97,35)	11587 (11587,0)	11593 (11593,0)
11597 (11597,0)	11617 (11617,0)	11621 (103,11)

11633 (11633,0)	11657 (11657,0)	11677 (11677,0)
11681 (98,31)	11689 (97,40)	11699 (99,26)
11701 (103,12)	11717 (11717,0)	11719 (97,42)
11731 (97,43)	11743 (11743,0)	11777 (11777,0)
11779 (103,13)	11783 (11783,0)	11789 (99,28)
11801 (105,8)	11807 (11807,0)	11813 (11813,0)
11821 (101,20)	11827 (11827,0)	11831 (99,29)
11833 (11833,0)	11839 (104,11)	11863 (11863,0)
11867 (11867,0)	11887 (11887,0)	11897 (11897,0)
11903 (11903,0)	11909 (99,31)	11923 (11923,0)
11927 (11927,0)	11933 (11933,0)	11939 (101,22)
11941 (98,41)	11953 (11953,0)	11959 (107,5)
11969 (98,43)	11971 (100,27)	11981 (102,19)
11987 (11987,0)	12007 (12007,0)	12011 (99,34)
12037 (12037,0)	12041 (99,35)	12043 (12043,0)
12049 (101,24)	12071 (103,17)	12073 (12073,0)
12097 (12097,0)	12101 (101,25)	12107 (12107,0)
12109 (106,9)	12113 (12113,0)	12119 (99,38)
12143 (12143,0)	12149 (107,7)	12157 (12157,0)
12161 (99,40)	12163 (12163,0)	12197 (12197,0)
12203 (12203,0)	12211 (100,33)	12227 (12227,0)
12239 (99,46)	12241 (107,8)	12251 (99,49)
12253 (12253,0)	12263 (12263,0)	12269 (103,20)
12277 (12277,0)	12281 (106,11)	12289 (101,29)
12301 (109,4)	12323 (12323,0)	12329 (102,25)
12343 (12343,0)	12347 (12347,0)	12373 (12373,0)
12377 (12377,0)	12379 (100,39)	12391 (103,22)
12401 (109,5)	12409 (101,32)	12413 (12413,0)
12421 (110,3)	12433 (12433,0)	12437 (12437,0)
12451 (100,43)	12457 (12457,0)	12473 (12473,0)
12479 (101,34)	12487 (12487,0)	12491 (100,47)
12497 (12497,0)	12503 (12503,0)	12511 (101,35)
12517 (12517,0)	12527 (12527,0)	12539 (111,2)
12541 (101,36)	12547 (12547,0)	12553 (12553,0)
12569 (101,37)	12577 (12577,0)	12583 (12583,0)
12589 (107,12)	12601 (106,15)	12611 (103,26)
12613 (12613,0)	12619 (101,39)	12637 (12637,0)
12641 (101,40)	12647 (12647,0)	12653 (12653,0)
12659 (105,19)	12671 (107,13)	12689 (109,8)
12697 (12697,0)	12703 (12703,0)	12713 (12713,0)
12721 (101,45)	12739 (101,47)	12743 (12743,0)
12757 (12757,0)	12763 (12763,0)	12781 (109,9)
12791 (104,25)	12799 (103,30)	12809 (102,37)
12821 (110,7)	12823 (12823,0)	12829 (107,15)
12841 (103,31)	12853 (12853,0)	12889 (106,19)
12893 (12893,0)	12899 (108,13)	12907 (12907,0)
12911 (105,23)	12917 (12917,0)	12919 (103,33)
12923 (12923,0)	12941 (102,43)	12953 (12953,0)
12959 (109,11)	12967 (12967,0)	12973 (12973,0)
12979 (107,17)	12983 (12983,0)	13001 (102,49)
13003 (13003,0)	13007 (13007,0)	13009 (110,9)
13033 (13033,0)	13037 (13037,0)	13043 (13043,0)
13049 (111,7)	13063 (13063,0)	13093 (13093,0)
13099 (113,3)	13103 (13103,0)	13109 (114,1)

13121 (107,19)	13127 (13127,0)	13147 (13147,0)
13151 (103,41)	13159 (104,33)	13163 (13163,0)
13171 (103,42)	13177 (13177,0)	13183 (13183,0)
13187 (13187,0)	13217 (13217,0)	13219 (103,45)
13229 (105,29)	13241 (103,47)	13249 (103,48)
13259 (103,50)	13267 (13267,0)	13291 (109,15)
13297 (13297,0)	13309 (113,5)	13313 (13313,0)
13327 (13327,0)	13331 (111,10)	13337 (13337,0)
13339 (115,1)	13367 (13367,0)	13381 (107,23)
13397 (13397,0)	13399 (104,41)	13411 (113,6)
13417 (13417,0)	13421 (111,11)	13441 (107,24)
13451 (115,2)	13457 (13457,0)	13463 (13463,0)
13469 (106,29)	13477 (13477,0)	13487 (13487,0)
13499 (107,25)	13513 (13513,0)	13523 (13523,0)
13537 (13537,0)	13553 (13553,0)	13567 (13567,0)
13577 (13577,0)	13591 (109,19)	13597 (13597,0)
13613 (13613,0)	13619 (108,23)	13627 (13627,0)
13633 (13633,0)	13649 (105,41)	13669 (115,4)
13679 (111,14)	13681 (110,17)	13687 (13687,0)
13691 (105,43)	13693 (13693,0)	13697 (13697,0)
13709 (105,44)	13711 (107,29)	13721 (106,35)
13723 (13723,0)	13729 (109,21)	13751 (105,47)
13757 (13757,0)	13759 (107,30)	13763 (13763,0)
13781 (105,52)	13789 (106,37)	13799 (113,10)
13807 (13807,0)	13829 (110,19)	13831 (112,13)
13841 (111,16)	13859 (109,23)	13873 (13873,0)
13877 (13877,0)	13879 (115,6)	13883 (13883,0)
13901 (106,41)	13903 (13903,0)	13907 (13907,0)
13913 (13913,0)	13921 (109,24)	13931 (107,34)
13933 (13933,0)	13963 (13963,0)	13967 (13967,0)
13997 (13997,0)	13999 (112,15)	14009 (106,47)
14011 (116,5)	14029 (106,49)	14033 (14033,0)
14051 (108,31)	14057 (14057,0)	14071 (107,38)
14081 (115,8)	14083 (14083,0)	14087 (14087,0)
14107 (14107,0)	14143 (14143,0)	14149 (109,28)
14153 (14153,0)	14159 (112,17)	14173 (14173,0)
14177 (14177,0)	14197 (14197,0)	14207 (14207,0)
14221 (107,44)	14243 (14243,0)	14249 (117,5)
14251 (109,30)	14281 (107,48)	14293 (14293,0)
14303 (14303,0)	14321 (113,16)	14323 (14323,0)
14327 (14327,0)	14341 (110,27)	14347 (14347,0)
14369 (115,11)	14387 (14387,0)	14389 (109,33)
14401 (113,17)	14407 (14407,0)	14411 (108,41)
14419 (116,9)	14423 (14423,0)	14431 (109,34)
14437 (14437,0)	14447 (14447,0)	14449 (110,29)
14461 (115,12)	14479 (113,18)	14489 (118,5)
14503 (14503,0)	14519 (120,1)	14533 (14533,0)
14537 (14537,0)	14543 (14543,0)	14549 (110,31)
14551 (115,13)	14557 (14557,0)	14561 (117,8)
14563 (14563,0)	14591 (112,23)	14593 (14593,0)
14621 (119,4)	14627 (14627,0)	14629 (113,20)
14633 (14633,0)	14639 (115,14)	14653 (14653,0)

14657 (14657,0)	14669 (109,41)	14683 (14683,0)
14699 (111,29)	14713 (14713,0)	14717 (14717,0)
14723 (14723,0)	14731 (119,5)	14737 (14737,0)
14741 (109,44)	14747 (14747,0)	14753 (14753,0)
14759 (117,10)	14767 (14767,0)	14771 (113,22)
14779 (109,46)	14783 (14783,0)	14797 (14797,0)
14813 (14813,0)	14821 (109,49)	14827 (14827,0)
14831 (109,50)	14843 (14843,0)	14851 (109,54)
14867 (14867,0)	14869 (110,39)	14879 (121,2)
14887 (14887,0)	14891 (115,17)	14897 (14897,0)
14923 (14923,0)	14929 (110,41)	14939 (111,34)
14947 (14947,0)	14951 (112,29)	14957 (14957,0)
14969 (113,25)	14983 (14983,0)	15013 (15013,0)
15017 (15017,0)	15031 (113,26)	15053 (15053,0)
15061 (110,47)	15073 (15073,0)	15077 (15077,0)
15083 (15083,0)	15091 (113,27)	15101 (118,11)
15107 (15107,0)	15121 (110,53)	15131 (117,14)
15137 (15137,0)	15139 (116,17)	15149 (113,28)
15161 (111,40)	15173 (15173,0)	15187 (15187,0)
15193 (15193,0)	15199 (115,21)	15217 (15217,0)
15227 (15227,0)	15233 (15233,0)	15241 (122,3)
15259 (113,30)	15263 (15263,0)	15269 (111,44)
15271 (115,22)	15277 (15277,0)	15287 (15287,0)
15289 (118,13)	15299 (116,19)	15307 (15307,0)
15313 (15313,0)	15319 (112,37)	15329 (111,47)
15331 (121,6)	15349 (119,11)	15359 (111,49)
15361 (113,32)	15373 (15373,0)	15377 (15377,0)
15383 (15383,0)	15391 (112,39)	15401 (111,55)
15413 (15413,0)	15427 (15427,0)	15439 (121,7)
15443 (15443,0)	15451 (116,21)	15461 (114,29)
15467 (15467,0)	15473 (15473,0)	15493 (15493,0)
15497 (15497,0)	15511 (112,43)	15527 (15527,0)
15541 (113,36)	15551 (117,19)	15559 (112,45)
15569 (114,31)	15581 (113,37)	15583 (15583,0)
15601 (115,27)	15607 (15607,0)	15619 (113,38)
15629 (117,20)	15641 (118,17)	15643 (15643,0)
15647 (15647,0)	15649 (121,9)	15661 (115,28)
15667 (15667,0)	15671 (112,53)	15679 (112,55)
15683 (15683,0)	15727 (15727,0)	15731 (116,25)
15733 (15733,0)	15737 (15737,0)	15739 (124,3)
15749 (125,1)	15761 (114,35)	15767 (15767,0)
15773 (15773,0)	15787 (15787,0)	15791 (120,13)
15797 (15797,0)	15803 (15803,0)	15809 (119,16)
15817 (15817,0)	15823 (15823,0)	15859 (116,27)
15877 (15877,0)	15881 (115,32)	15887 (15887,0)
15889 (113,48)	15901 (122,9)	15907 (15907,0)
15913 (15913,0)	15919 (113,50)	15923 (15923,0)
15937 (15937,0)	15959 (113,55)	15971 (124,5)
15973 (15973,0)	15991 (125,3)	16001 (126,1)
16007 (16007,0)	16033 (16033,0)	16057 (16057,0)
16061 (119,19)	16063 (16063,0)	16067 (16067,0)
16069 (115,36)	16073 (16073,0)	16087 (16087,0)

16091 (116,31)	16097 (16097,0)	16103 (16103,0)
16111 (115,37)	16127 (16127,0)	16139 (121,14)
16141 (119,20)	16183 (16183,0)	16187 (16187,0)
16189 (115,39)	16193 (16193,0)	16217 (16217,0)
16223 (16223,0)	16229 (114,53)	16231 (121,15)
16249 (118,25)	16253 (16253,0)	16267 (16267,0)
16273 (16273,0)	16301 (122,13)	16319 (120,19)
16333 (16333,0)	16339 (125,6)	16349 (115,44)
16361 (123,11)	16363 (16363,0)	16369 (119,23)
16381 (118,27)	16411 (124,9)	16417 (16417,0)
16421 (115,47)	16427 (16427,0)	16433 (16433,0)
16447 (16447,0)	16451 (125,7)	16453 (16453,0)
16477 (16477,0)	16481 (126,5)	16487 (16487,0)
16493 (16493,0)	16519 (115,54)	16529 (115,56)
16547 (16547,0)	16553 (16553,0)	16561 (125,8)
16567 (16567,0)	16573 (16573,0)	16603 (16603,0)
16607 (16607,0)	16619 (124,11)	16631 (120,23)
16633 (16633,0)	16649 (117,37)	16651 (116,45)
16657 (16657,0)	16661 (121,20)	16673 (16673,0)
16691 (117,38)	16693 (16693,0)	16699 (116,47)
16703 (16703,0)	16729 (118,33)	16741 (121,21)
16747 (16747,0)	16759 (128,3)	16763 (16763,0)
16787 (16787,0)	16811 (116,55)	16823 (16823,0)
16829 (118,35)	16831 (119,30)	16843 (16843,0)
16871 (117,43)	16879 (125,11)	16883 (16883,0)
16889 (119,31)	16901 (117,44)	16903 (16903,0)
16921 (118,37)	16927 (16927,0)	16931 (123,17)
16937 (16937,0)	16943 (16943,0)	16963 (16963,0)
16979 (117,47)	16981 (125,12)	16987 (16987,0)
16993 (16993,0)	17011 (124,15)	17021 (117,49)
17027 (17027,0)	17029 (130,1)	17033 (17033,0)
17041 (121,25)	17047 (17047,0)	17053 (17053,0)
17077 (17077,0)	17093 (17093,0)	17099 (117,55)
17107 (17107,0)	17117 (17117,0)	17123 (17123,0)
17137 (17137,0)	17159 (120,31)	17167 (17167,0)
17183 (17183,0)	17189 (123,20)	17191 (127,9)
17203 (17203,0)	17207 (17207,0)	17209 (118,45)
17231 (128,7)	17239 (119,38)	17257 (17257,0)
17291 (131,1)	17293 (17293,0)	17299 (127,10)
17317 (17317,0)	17321 (119,40)	17327 (17327,0)
17333 (17333,0)	17341 (118,51)	17351 (123,22)
17359 (119,41)	17377 (17377,0)	17383 (17383,0)
17387 (17387,0)	17389 (118,55)	17393 (17393,0)
17401 (118,57)	17417 (17417,0)	17419 (131,2)
17431 (121,31)	17443 (17443,0)	17449 (122,27)
17467 (17467,0)	17471 (120,37)	17477 (17477,0)
17483 (17483,0)	17489 (121,32)	17491 (119,45)
17497 (17497,0)	17509 (127,12)	17519 (119,46)
17539 (124,21)	17551 (125,18)	17569 (119,48)
17573 (17573,0)	17579 (123,25)	17581 (122,29)
17597 (17597,0)	17599 (121,34)	17609 (129,8)
17623 (17623,0)	17627 (17627,0)	17657 (17657,0)

17659 (119,53)	17669 (131,4)	17681 (119,55)
17683 (17683,0)	17707 (17707,0)	17713 (17713,0)
17729 (126,17)	17737 (17737,0)	17747 (17747,0)
17749 (121,37)	17761 (130,7)	17783 (17783,0)
17789 (123,28)	17791 (131,5)	17807 (17807,0)
17827 (17827,0)	17837 (17837,0)	17839 (121,39)
17851 (124,25)	17863 (17863,0)	17881 (121,40)
17891 (125,22)	17903 (17903,0)	17909 (126,19)
17911 (131,6)	17921 (121,41)	17923 (17923,0)
17929 (122,35)	17939 (129,11)	17957 (17957,0)
17959 (121,42)	17971 (125,23)	17977 (17977,0)
17981 (123,31)	17987 (17987,0)	17989 (130,9)
18013 (18013,0)	18041 (123,32)	18043 (18043,0)
18047 (18047,0)	18049 (125,24)	18059 (132,5)
18061 (121,45)	18077 (18077,0)	18089 (134,1)
18097 (18097,0)	18119 (121,47)	18121 (122,39)
18127 (18127,0)	18131 (124,29)	18133 (18133,0)
18143 (18143,0)	18149 (129,13)	18169 (121,49)
18181 (127,19)	18191 (121,50)	18199 (125,26)
18211 (121,51)	18217 (18217,0)	18223 (18223,0)
18229 (121,52)	18233 (18233,0)	18251 (129,14)
18253 (18253,0)	18257 (18257,0)	18269 (127,20)
18287 (18287,0)	18289 (121,57)	18301 (121,60)
18307 (18307,0)	18311 (123,37)	18313 (18313,0)
18329 (133,5)	18341 (125,28)	18353 (18353,0)
18367 (18367,0)	18371 (131,10)	18379 (124,33)
18397 (18397,0)	18401 (126,25)	18413 (18413,0)
18427 (18427,0)	18433 (18433,0)	18439 (127,22)
18443 (18443,0)	18451 (133,6)	18457 (18457,0)
18461 (122,49)	18481 (131,11)	18493 (18493,0)
18503 (18503,0)	18517 (18517,0)	18521 (127,23)
18523 (18523,0)	18539 (125,31)	18541 (122,53)
18553 (18553,0)	18583 (18583,0)	18587 (18587,0)
18593 (18593,0)	18617 (18617,0)	18637 (18637,0)
18661 (125,33)	18671 (123,46)	18679 (127,25)
18691 (124,39)	18701 (123,47)	18713 (18713,0)
18719 (125,34)	18731 (129,19)	18743 (18743,0)
18749 (135,4)	18757 (18757,0)	18773 (18773,0)
18787 (18787,0)	18793 (18793,0)	18797 (18797,0)
18803 (18803,0)	18839 (123,53)	18859 (124,43)
18869 (123,55)	18899 (123,58)	18911 (123,61)
18913 (18913,0)	18917 (18917,0)	18919 (133,10)
18947 (18947,0)	18959 (128,25)	18973 (18973,0)
18979 (125,39)	19001 (131,16)	19009 (130,19)
19013 (19013,0)	19031 (133,11)	19037 (19037,0)
19051 (124,49)	19069 (125,41)	19073 (19073,0)
19079 (129,23)	19081 (134,9)	19087 (19087,0)
19121 (135,7)	19139 (124,53)	19141 (133,12)
19157 (19157,0)	19163 (19163,0)	19181 (138,1)
19183 (19183,0)	19207 (19207,0)	19211 (124,59)
19213 (19213,0)	19219 (124,61)	19231 (127,33)
19237 (19237,0)	19249 (133,13)	19259 (125,46)

19267 (19267,0)	19273 (19273,0)	19289 (131,19)
19301 (137,4)	19309 (134,11)	19319 (129,26)
19333 (19333,0)	19373 (19373,0)	19379 (132,17)
19381 (131,20)	19387 (19387,0)	19391 (128,31)
19403 (19403,0)	19417 (19417,0)	19421 (125,52)
19423 (19423,0)	19427 (19427,0)	19429 (137,5)
19433 (19433,0)	19441 (125,53)	19447 (19447,0)
19457 (19457,0)	19463 (19463,0)	19469 (129,28)
19471 (131,21)	19477 (19477,0)	19483 (19483,0)
19489 (125,56)	19501 (125,57)	19507 (19507,0)
19531 (125,62)	19541 (129,29)	19543 (19543,0)
19553 (19553,0)	19559 (131,22)	19571 (132,19)
19577 (19577,0)	19583 (19583,0)	19597 (19597,0)
19603 (19603,0)	19609 (127,40)	19661 (133,17)
19681 (130,27)	19687 (19687,0)	19697 (19697,0)
19699 (127,42)	19709 (138,5)	19717 (19717,0)
19727 (19727,0)	19739 (140,1)	19751 (128,37)
19753 (19753,0)	19759 (133,18)	19763 (19763,0)
19777 (19777,0)	19793 (19793,0)	19801 (137,8)
19813 (19813,0)	19819 (127,45)	19841 (126,61)
19843 (19843,0)	19853 (19853,0)	19861 (139,4)
19867 (19867,0)	19889 (127,47)	19891 (131,26)
19913 (19913,0)	19919 (135,14)	19927 (19927,0)
19937 (19937,0)	19949 (133,20)	19961 (138,7)
19963 (19963,0)	19973 (19973,0)	19979 (127,50)
19991 (139,5)	19993 (19993,0)	19997 (19997,0)
20011 (140,3)	20021 (141,1)	20023 (20023,0)
20029 (127,52)	20047 (20047,0)	20051 (127,53)
20063 (20063,0)	20071 (127,54)	20089 (127,55)
20101 (130,33)	20107 (20107,0)	20113 (20113,0)
20117 (20117,0)	20123 (20123,0)	20129 (135,16)
20143 (20143,0)	20147 (20147,0)	20149 (127,60)
20161 (127,63)	20173 (20173,0)	20177 (20177,0)
20183 (20183,0)	20201 (129,40)	20219 (133,23)
20231 (135,17)	20233 (20233,0)	20249 (129,41)
20261 (131,31)	20269 (137,12)	20287 (20287,0)
20297 (20297,0)	20323 (20323,0)	20327 (20327,0)
20333 (20333,0)	20341 (130,37)	20347 (20347,0)
20353 (20353,0)	20357 (20357,0)	20359 (128,53)
20369 (139,8)	20389 (133,25)	20393 (20393,0)
20399 (128,55)	20407 (20407,0)	20411 (132,29)
20431 (128,57)	20441 (138,11)	20443 (20443,0)
20477 (20477,0)	20479 (128,63)	20483 (20483,0)
20507 (20507,0)	20509 (134,23)	20521 (131,35)
20533 (20533,0)	20543 (20543,0)	20549 (130,41)
20551 (133,27)	20563 (20563,0)	20593 (20593,0)
20599 (137,15)	20611 (139,10)	20627 (20627,0)
20639 (131,37)	20641 (130,43)	20663 (20663,0)
20681 (134,25)	20693 (20693,0)	20707 (20707,0)
20717 (20717,0)	20719 (136,19)	20731 (143,2)
20743 (20743,0)	20747 (20747,0)	20749 (131,39)
20753 (20753,0)	20759 (129,58)	20771 (129,59)

20773 (20773,0)	20789 (129,61)	20807 (20807,0)
20809 (137,17)	20849 (142,5)	20857 (20857,0)
20873 (20873,0)	20879 (144,1)	20887 (20887,0)
20897 (20897,0)	20899 (131,42)	20903 (20903,0)
20921 (133,32)	20929 (130,51)	20939 (132,37)
20947 (20947,0)	20959 (139,13)	20963 (20963,0)
20981 (130,53)	20983 (20983,0)	21001 (134,29)
21011 (137,19)	21013 (21013,0)	21017 (21017,0)
21019 (140,11)	21023 (21023,0)	21031 (131,45)
21059 (135,26)	21061 (130,57)	21067 (21067,0)
21089 (130,59)	21101 (138,17)	21107 (21107,0)
21121 (130,63)	21139 (143,5)	21143 (21143,0)
21149 (134,31)	21157 (21157,0)	21163 (21163,0)
21169 (145,1)	21179 (131,49)	21187 (21187,0)
21191 (141,10)	21193 (21193,0)	21211 (131,50)
21221 (135,28)	21227 (21227,0)	21247 (21247,0)
21269 (131,52)	21277 (21277,0)	21283 (21283,0)
21313 (21313,0)	21317 (21317,0)	21319 (131,54)
21323 (21323,0)	21341 (131,55)	21347 (21347,0)
21377 (21377,0)	21379 (131,57)	21383 (21383,0)
21391 (137,23)	21397 (21397,0)	21401 (143,7)
21407 (21407,0)	21419 (132,47)	21433 (21433,0)
21467 (21467,0)	21481 (137,24)	21487 (21487,0)
21491 (132,49)	21493 (21493,0)	21499 (139,18)
21503 (21503,0)	21517 (21517,0)	21521 (135,32)
21523 (21523,0)	21529 (143,8)	21557 (21557,0)
21559 (133,43)	21563 (21563,0)	21569 (137,25)
21577 (21577,0)	21587 (21587,0)	21589 (145,4)
21599 (136,29)	21601 (139,19)	21611 (132,53)
21613 (21613,0)	21617 (21617,0)	21647 (21647,0)
21649 (133,45)	21661 (134,39)	21673 (21673,0)
21683 (21683,0)	21701 (139,20)	21713 (21713,0)
21727 (21727,0)	21737 (21737,0)	21739 (137,27)
21751 (136,31)	21757 (21757,0)	21767 (21767,0)
21773 (21773,0)	21787 (21787,0)	21799 (139,21)
21803 (21803,0)	21817 (21817,0)	21821 (137,28)
21839 (133,50)	21841 (142,13)	21851 (135,37)
21859 (145,6)	21863 (21863,0)	21871 (133,51)
21881 (141,16)	21893 (21893,0)	21911 (135,38)
21929 (133,53)	21937 (21937,0)	21943 (21943,0)
21961 (134,45)	21977 (21977,0)	21991 (145,7)
21997 (21997,0)	22003 (22003,0)	22013 (22013,0)
22027 (22027,0)	22031 (136,35)	22037 (22037,0)
22039 (133,58)	22051 (148,1)	22063 (22063,0)
22067 (22067,0)	22073 (22073,0)	22079 (135,41)
22091 (133,62)	22093 (22093,0)	22109 (133,65)
22111 (133,66)	22123 (22123,0)	22129 (137,32)
22133 (22133,0)	22147 (22147,0)	22153 (22153,0)
22157 (22157,0)	22159 (136,37)	22171 (139,25)
22189 (134,51)	22193 (22193,0)	22229 (135,44)
22247 (22247,0)	22259 (139,26)	22271 (137,34)
22273 (22273,0)	22277 (22277,0)	22279 (136,39)

22283 (22283,0)	22291 (140,23)	22303 (22303,0)
22307 (22307,0)	22343 (22343,0)	22349 (149,1)
22367 (22367,0)	22369 (143,15)	22381 (134,59)
22391 (136,41)	22397 (22397,0)	22409 (134,61)
22433 (22433,0)	22441 (134,65)	22447 (22447,0)
22453 (22453,0)	22469 (137,37)	22481 (143,16)
22483 (22483,0)	22501 (142,19)	22511 (139,29)
22531 (137,38)	22541 (135,52)	22543 (22543,0)
22549 (146,9)	22567 (22567,0)	22571 (135,53)
22573 (22573,0)	22613 (22613,0)	22619 (148,5)
22621 (145,12)	22637 (22637,0)	22639 (149,3)
22643 (22643,0)	22651 (140,27)	22669 (139,31)
22679 (136,47)	22691 (135,58)	22697 (22697,0)
22699 (143,18)	22709 (135,59)	22717 (22717,0)
22721 (147,8)	22727 (22727,0)	22739 (135,61)
22741 (145,13)	22751 (135,62)	22769 (135,64)
22777 (22777,0)	22783 (22783,0)	22787 (22787,0)
22807 (22807,0)	22811 (137,43)	22817 (22817,0)
22853 (22853,0)	22859 (145,14)	22861 (137,44)
22871 (141,26)	22877 (22877,0)	22901 (142,23)
22907 (22907,0)	22921 (149,5)	22937 (22937,0)
22943 (22943,0)	22961 (139,35)	22963 (22963,0)
22973 (22973,0)	22993 (22993,0)	23003 (23003,0)
23011 (143,21)	23017 (23017,0)	23021 (138,41)
23027 (23027,0)	23029 (139,36)	23039 (136,59)
23041 (137,48)	23053 (23053,0)	23057 (23057,0)
23059 (149,6)	23063 (23063,0)	23071 (136,61)
23081 (137,49)	23087 (23087,0)	23099 (151,2)
23117 (23117,0)	23131 (140,33)	23143 (23143,0)
23159 (139,38)	23167 (23167,0)	23173 (23173,0)
23189 (137,52)	23197 (23197,0)	23201 (145,17)
23203 (23203,0)	23209 (143,23)	23227 (23227,0)
23251 (137,54)	23269 (142,27)	23279 (137,55)
23291 (141,31)	23293 (23293,0)	23297 (23297,0)
23311 (145,18)	23321 (138,47)	23327 (23327,0)
23333 (23333,0)	23339 (139,41)	23357 (23357,0)
23369 (141,32)	23371 (137,59)	23399 (143,25)
23417 (23417,0)	23431 (137,63)	23447 (23447,0)
23459 (137,67)	23473 (23473,0)	23497 (23497,0)
23509 (146,17)	23531 (151,5)	23537 (23537,0)
23539 (140,39)	23549 (138,53)	23557 (23557,0)
23561 (153,1)	23563 (23563,0)	23567 (23567,0)
23581 (143,27)	23593 (23593,0)	23599 (139,46)
23603 (23603,0)	23609 (138,55)	23623 (23623,0)
23627 (23627,0)	23629 (145,21)	23633 (23633,0)
23663 (23663,0)	23669 (143,28)	23671 (151,6)
23677 (23677,0)	23687 (23687,0)	23689 (139,48)
23719 (149,11)	23741 (138,61)	23743 (23743,0)
23747 (23747,0)	23753 (23753,0)	23761 (142,33)
23767 (23767,0)	23773 (23773,0)	23789 (138,65)
23801 (138,67)	23813 (23813,0)	23819 (147,17)
23827 (23827,0)	23831 (145,23)	23833 (23833,0)

23857 (23857,0)	23869 (154,1)	23873 (23873,0)
23879 (139,53)	23887 (23887,0)	23893 (23893,0)
23899 (148,15)	23909 (142,35)	23911 (139,54)
23917 (23917,0)	23929 (145,24)	23957 (23957,0)
23971 (140,47)	23977 (23977,0)	23981 (141,41)
23993 (23993,0)	24001 (143,32)	24007 (24007,0)
24019 (139,58)	24023 (24023,0)	24029 (150,11)
24043 (24043,0)	24049 (142,37)	24061 (139,60)
24071 (144,29)	24077 (24077,0)	24083 (24083,0)
24091 (149,14)	24097 (24097,0)	24103 (24103,0)
24107 (24107,0)	24109 (139,63)	24113 (24113,0)
24121 (139,64)	24133 (24133,0)	24137 (24137,0)
24151 (139,69)	24169 (154,3)	24179 (155,1)
24181 (142,39)	24197 (24197,0)	24203 (24203,0)
24223 (24223,0)	24229 (143,35)	24239 (144,31)
24247 (24247,0)	24251 (141,46)	24281 (150,13)
24317 (24317,0)	24329 (149,16)	24337 (24337,0)
24359 (147,22)	24371 (143,37)	24373 (24373,0)
24379 (140,59)	24391 (152,9)	24407 (24407,0)
24413 (24413,0)	24419 (140,61)	24421 (142,43)
24439 (143,38)	24443 (24443,0)	24469 (151,12)
24473 (24473,0)	24481 (155,3)	24499 (140,69)
24509 (141,52)	24517 (24517,0)	24527 (24527,0)
24533 (24533,0)	24547 (24547,0)	24551 (144,35)
24571 (148,21)	24593 (24593,0)	24611 (141,55)
24623 (24623,0)	24631 (143,41)	24659 (147,25)
24671 (149,19)	24677 (24677,0)	24683 (24683,0)
24691 (143,42)	24697 (24697,0)	24709 (146,29)
24733 (24733,0)	24749 (143,43)	24763 (24763,0)
24767 (24767,0)	24781 (149,20)	24793 (24793,0)
24799 (145,34)	24809 (141,64)	24821 (141,65)
24841 (151,15)	24847 (24847,0)	24851 (141,70)
24859 (143,45)	24877 (24877,0)	24889 (149,21)
24907 (24907,0)	24917 (24917,0)	24919 (155,6)
24923 (24923,0)	24943 (24943,0)	24953 (24953,0)
24967 (24967,0)	24971 (153,11)	24977 (24977,0)
24979 (148,25)	24989 (150,19)	25013 (25013,0)
25031 (147,29)	25033 (25033,0)	25037 (25037,0)
25057 (25057,0)	25073 (25073,0)	25087 (25087,0)
25097 (25097,0)	25111 (157,3)	25117 (25117,0)
25121 (158,1)	25127 (25127,0)	25147 (25147,0)
25153 (25153,0)	25163 (25163,0)	25169 (142,65)
25171 (148,27)	25183 (25183,0)	25189 (142,67)
25219 (143,53)	25229 (153,13)	25237 (25237,0)
25243 (25243,0)	25247 (25247,0)	25253 (25253,0)
25261 (157,4)	25301 (149,25)	25303 (25303,0)
25307 (25307,0)	25309 (151,19)	25321 (143,56)
25339 (155,9)	25343 (25343,0)	25349 (146,37)
25357 (25357,0)	25367 (25367,0)	25373 (25373,0)
25391 (144,49)	25409 (157,5)	25411 (145,43)
25423 (25423,0)	25439 (159,1)	25447 (25447,0)
25453 (25453,0)	25457 (25457,0)	25463 (25463,0)

25469 (145,44)	25471 (143,62)	25523 (25523,0)
25537 (25537,0)	25541 (143,67)	25561 (143,71)
25577 (25577,0)	25579 (145,46)	25583 (25583,0)
25589 (149,28)	25601 (153,16)	25603 (25603,0)
25609 (155,11)	25621 (146,41)	25633 (25633,0)
25639 (151,22)	25643 (25643,0)	25657 (25657,0)
25667 (25667,0)	25673 (25673,0)	25679 (147,37)
25693 (25693,0)	25703 (25703,0)	25717 (25717,0)
25733 (25733,0)	25741 (155,12)	25747 (25747,0)
25759 (160,1)	25763 (25763,0)	25771 (149,30)
25793 (25793,0)	25799 (144,61)	25801 (154,15)
25819 (145,51)	25841 (157,8)	25847 (25847,0)
25849 (151,24)	25867 (25867,0)	25873 (25873,0)
25889 (147,40)	25903 (25903,0)	25913 (25913,0)
25919 (144,71)	25931 (156,11)	25933 (25933,0)
25939 (145,54)	25943 (25943,0)	25951 (151,25)
25969 (146,47)	25981 (157,9)	25997 (25997,0)
25999 (155,14)	26003 (26003,0)	26017 (26017,0)
26021 (158,7)	26029 (149,33)	26041 (145,57)
26053 (26053,0)	26083 (26083,0)	26099 (145,59)
26107 (26107,0)	26111 (149,34)	26113 (26113,0)
26119 (157,10)	26141 (147,44)	26153 (26153,0)
26161 (146,51)	26171 (145,62)	26177 (26177,0)
26183 (26183,0)	26189 (150,31)	26203 (26203,0)
26209 (145,64)	26227 (26227,0)	26237 (26237,0)
26249 (155,16)	26251 (145,67)	26261 (145,68)
26263 (26263,0)	26267 (26267,0)	26293 (26293,0)
26297 (26297,0)	26309 (147,47)	26317 (26317,0)
26321 (146,55)	26339 (151,29)	26347 (26347,0)
26357 (26357,0)	26371 (155,17)	26387 (26387,0)
26393 (26393,0)	26399 (153,23)	26407 (26407,0)
26417 (26417,0)	26423 (26423,0)	26431 (151,30)
26437 (26437,0)	26449 (146,59)	26459 (147,50)
26479 (152,27)	26489 (159,8)	26497 (26497,0)
26501 (146,61)	26513 (26513,0)	26539 (148,45)
26557 (26557,0)	26561 (149,40)	26573 (26573,0)
26591 (147,53)	26597 (26597,0)	26627 (26627,0)
26633 (26633,0)	26641 (146,71)	26647 (26647,0)
26669 (147,55)	26681 (150,37)	26683 (26683,0)
26687 (26687,0)	26693 (26693,0)	26699 (156,17)
26701 (161,5)	26711 (153,26)	26713 (26713,0)
26717 (26717,0)	26723 (26723,0)	26729 (154,23)
26731 (163,1)	26737 (26737,0)	26759 (149,43)
26777 (26777,0)	26783 (26783,0)	26801 (147,59)
26813 (26813,0)	26821 (149,44)	26833 (26833,0)
26839 (155,21)	26849 (158,13)	26861 (151,35)
26863 (26863,0)	26879 (147,62)	26881 (149,45)
26891 (163,2)	26893 (26893,0)	26903 (26903,0)
26921 (147,64)	26927 (26927,0)	26947 (26947,0)
26951 (155,22)	26953 (26953,0)	26959 (160,9)
26981 (147,68)	26987 (26987,0)	26993 (26993,0)
27011 (147,73)	27017 (27017,0)	27031 (152,33)

27043 (27043,0)	27059 (164,1)	27061 (155,23)
27067 (27067,0)	27073 (27073,0)	27077 (27077,0)
27091 (148,57)	27103 (27103,0)	27107 (27107,0)
27109 (158,15)	27127 (27127,0)	27143 (27143,0)
27179 (159,13)	27191 (153,31)	27197 (27197,0)
27211 (148,61)	27239 (160,11)	27241 (151,40)
27253 (27253,0)	27259 (148,63)	27271 (157,19)
27277 (27277,0)	27281 (153,32)	27283 (27283,0)
27299 (148,65)	27329 (162,7)	27337 (27337,0)
27361 (158,17)	27367 (27367,0)	27397 (27397,0)
27407 (27407,0)	27409 (149,56)	27427 (27427,0)
27431 (161,10)	27437 (27437,0)	27449 (150,49)
27457 (27457,0)	27479 (149,58)	27481 (155,27)
27487 (27487,0)	27509 (151,44)	27527 (27527,0)
27529 (154,31)	27539 (153,35)	27541 (149,60)
27551 (165,2)	27581 (155,28)	27583 (27583,0)
27611 (156,25)	27617 (27617,0)	27631 (151,46)
27647 (27647,0)	27653 (27653,0)	27673 (27673,0)
27689 (151,47)	27691 (164,5)	27697 (27697,0)
27701 (153,37)	27733 (27733,0)	27737 (27737,0)
27739 (149,71)	27743 (27743,0)	27749 (149,73)
27751 (149,74)	27763 (27763,0)	27767 (27767,0)
27773 (27773,0)	27779 (153,38)	27791 (152,43)
27793 (27793,0)	27799 (151,49)	27803 (27803,0)
27809 (163,8)	27817 (27817,0)	27823 (27823,0)
27827 (27827,0)	27847 (27847,0)	27851 (151,50)
27883 (27883,0)	27893 (27893,0)	27901 (151,51)
27917 (27917,0)	27919 (152,45)	27941 (159,19)
27943 (27943,0)	27947 (27947,0)	27953 (27953,0)
27961 (155,32)	27967 (27967,0)	27983 (27983,0)
27997 (27997,0)	28001 (153,41)	28019 (156,29)
28027 (28027,0)	28031 (160,17)	28051 (155,33)
28057 (28057,0)	28069 (158,23)	28081 (151,55)
28087 (28087,0)	28097 (28097,0)	28099 (163,10)
28109 (150,71)	28111 (161,15)	28123 (28123,0)
28151 (152,49)	28163 (28163,0)	28181 (162,13)
28183 (28183,0)	28201 (154,39)	28211 (156,31)
28219 (167,2)	28229 (151,59)	28277 (28277,0)
28279 (160,19)	28283 (28283,0)	28289 (158,25)
28297 (28297,0)	28307 (28307,0)	28309 (155,36)
28319 (151,62)	28349 (154,41)	28351 (152,53)
28387 (28387,0)	28393 (28393,0)	28403 (28403,0)
28409 (159,23)	28411 (151,66)	28429 (151,67)
28433 (28433,0)	28439 (152,55)	28447 (28447,0)
28463 (28463,0)	28477 (28477,0)	28493 (28493,0)
28499 (151,74)	28513 (28513,0)	28517 (28517,0)
28537 (28537,0)	28541 (167,4)	28547 (28547,0)
28549 (155,39)	28559 (153,50)	28571 (156,35)
28573 (28573,0)	28579 (164,11)	28591 (152,59)
28597 (28597,0)	28603 (28603,0)	28607 (28607,0)
28619 (161,19)	28621 (154,45)	28627 (28627,0)
28631 (159,25)	28643 (28643,0)	28649 (157,32)

28657 (28657,0)	28661 (153,52)	28663 (28663,0)
28669 (166,7)	28687 (28687,0)	28697 (28697,0)
28703 (28703,0)	28711 (152,63)	28723 (28723,0)
28729 (169,1)	28751 (160,23)	28753 (28753,0)
28759 (152,65)	28771 (155,42)	28789 (163,15)
28793 (28793,0)	28807 (28807,0)	28813 (28813,0)
28817 (28817,0)	28837 (28837,0)	28843 (28843,0)
28859 (164,13)	28867 (28867,0)	28871 (152,73)
28879 (152,75)	28901 (158,31)	28909 (155,44)
28921 (163,16)	28927 (28927,0)	28933 (28933,0)
28949 (159,28)	28961 (162,19)	28979 (161,22)
29009 (167,7)	29017 (29017,0)	29021 (153,61)
29023 (29023,0)	29027 (29027,0)	29033 (29033,0)
29059 (169,3)	29063 (29063,0)	29077 (29077,0)
29101 (155,47)	29123 (29123,0)	29129 (153,65)
29131 (164,15)	29137 (29137,0)	29147 (29147,0)
29153 (29153,0)	29167 (29167,0)	29173 (29173,0)
29179 (163,18)	29191 (160,27)	29201 (165,13)
29207 (29207,0)	29209 (161,24)	29221 (169,4)
29231 (153,71)	29243 (29243,0)	29251 (157,39)
29269 (158,35)	29287 (29287,0)	29297 (29297,0)
29303 (29303,0)	29311 (167,9)	29327 (29327,0)
29333 (29333,0)	29339 (165,14)	29347 (29347,0)
29363 (29363,0)	29383 (29383,0)	29387 (29387,0)
29389 (154,61)	29399 (160,29)	29401 (170,3)
29411 (171,1)	29423 (29423,0)	29429 (163,20)
29437 (29437,0)	29443 (29443,0)	29453 (29453,0)
29473 (29473,0)	29483 (29483,0)	29501 (154,65)
29527 (29527,0)	29531 (159,34)	29537 (29537,0)
29567 (29567,0)	29569 (155,56)	29573 (29573,0)
29581 (154,69)	29587 (29587,0)	29599 (160,31)
29611 (155,57)	29629 (154,73)	29633 (29633,0)
29641 (154,75)	29663 (29663,0)	29669 (162,25)
29671 (163,22)	29683 (29683,0)	29717 (29717,0)
29723 (29723,0)	29741 (165,17)	29753 (29753,0)
29759 (155,61)	29761 (158,41)	29789 (163,23)
29803 (29803,0)	29819 (157,47)	29833 (29833,0)
29837 (29837,0)	29851 (161,30)	29863 (29863,0)
29867 (29867,0)	29873 (29873,0)	29879 (159,38)
29881 (157,48)	29917 (29917,0)	29921 (155,67)
29927 (29927,0)	29947 (29947,0)	29959 (155,69)
29983 (29983,0)	29989 (155,71)	30011 (155,73)
30013 (30013,0)	30029 (155,76)	30047 (30047,0)
30059 (156,59)	30071 (171,5)	30089 (166,17)
30091 (172,3)	30097 (30097,0)	30103 (30103,0)
30109 (157,52)	30113 (30113,0)	30119 (159,41)
30133 (30133,0)	30137 (30137,0)	30139 (164,23)
30161 (157,53)	30169 (167,15)	30181 (158,47)
30187 (30187,0)	30197 (30197,0)	30203 (30203,0)
30211 (157,54)	30223 (30223,0)	30241 (163,27)
30253 (30253,0)	30259 (157,55)	30269 (159,43)
30271 (173,2)	30293 (30293,0)	30307 (30307,0)

30313 (30313,0)	30319 (160,39)	30323 (30323,0)
30341 (159,44)	30347 (30347,0)	30367 (30367,0)
30389 (171,7)	30391 (157,58)	30403 (30403,0)
30427 (30427,0)	30431 (157,59)	30449 (174,1)
30467 (30467,0)	30469 (157,60)	30491 (165,23)
30493 (30493,0)	30497 (30497,0)	30509 (161,37)
30517 (30517,0)	30529 (158,53)	30539 (157,62)
30553 (30553,0)	30557 (30557,0)	30559 (163,30)
30577 (30577,0)	30593 (30593,0)	30631 (160,43)
30637 (30637,0)	30643 (30643,0)	30649 (170,11)
30661 (163,31)	30671 (159,49)	30677 (30677,0)
30689 (162,35)	30697 (30697,0)	30703 (30703,0)
30707 (30707,0)	30713 (30713,0)	30727 (30727,0)
30757 (30757,0)	30763 (30763,0)	30773 (30773,0)
30781 (157,73)	30803 (30803,0)	30809 (157,77)
30817 (30817,0)	30829 (167,20)	30839 (165,26)
30841 (161,41)	30851 (171,10)	30853 (30853,0)
30859 (163,33)	30869 (162,37)	30871 (169,15)
30881 (158,61)	30893 (30893,0)	30911 (160,47)
30931 (173,6)	30937 (30937,0)	30941 (170,13)
30949 (158,63)	30971 (175,2)	30977 (30977,0)
30983 (30983,0)	31013 (31013,0)	31019 (164,31)
31033 (31033,0)	31039 (160,49)	31051 (172,9)
31063 (31063,0)	31069 (161,44)	31079 (167,22)
31081 (166,25)	31091 (173,7)	31121 (174,5)
31123 (31123,0)	31139 (159,58)	31147 (31147,0)
31151 (176,1)	31153 (31153,0)	31159 (160,51)
31177 (31177,0)	31181 (159,59)	31183 (31183,0)
31189 (158,75)	31193 (31193,0)	31219 (164,33)
31223 (31223,0)	31231 (163,37)	31237 (31237,0)
31247 (31247,0)	31249 (173,8)	31253 (31253,0)
31259 (159,61)	31267 (31267,0)	31271 (160,53)
31277 (31277,0)	31307 (31307,0)	31319 (163,38)
31321 (167,24)	31327 (31327,0)	31333 (31333,0)
31337 (31337,0)	31357 (31357,0)	31379 (165,31)
31387 (31387,0)	31391 (159,65)	31393 (31393,0)
31397 (31397,0)	31469 (159,68)	31477 (31477,0)
31481 (165,32)	31489 (163,40)	31511 (159,70)
31513 (31513,0)	31517 (31517,0)	31531 (161,51)
31541 (169,20)	31543 (31543,0)	31547 (31547,0)
31567 (31567,0)	31573 (31573,0)	31583 (31583,0)
31601 (159,79)	31607 (31607,0)	31627 (31627,0)
31643 (31643,0)	31649 (162,47)	31657 (31657,0)
31663 (31663,0)	31667 (31667,0)	31687 (31687,0)
31699 (161,54)	31721 (171,16)	31723 (31723,0)
31727 (31727,0)	31729 (163,43)	31741 (166,31)
31751 (161,55)	31769 (170,19)	31771 (164,39)
31793 (31793,0)	31799 (168,25)	31817 (31817,0)
31847 (31847,0)	31849 (161,57)	31859 (171,17)
31873 (31873,0)	31883 (31883,0)	31891 (167,29)
31907 (31907,0)	31957 (31957,0)	31963 (31963,0)
31973 (31973,0)	31981 (161,60)	31991 (160,77)

32003 (32003,0)	32009 (173,13)	32027 (32027,0)
32029 (170,21)	32051 (165,38)	32057 (32057,0)
32059 (161,62)	32063 (32063,0)	32069 (174,11)
32077 (32077,0)	32083 (32083,0)	32089 (163,48)
32099 (164,43)	32117 (32117,0)	32119 (175,9)
32141 (166,35)	32143 (32143,0)	32159 (176,7)
32173 (32173,0)	32183 (32183,0)	32189 (177,5)
32191 (161,66)	32203 (32203,0)	32213 (32213,0)
32233 (32233,0)	32237 (32237,0)	32251 (164,45)
32257 (32257,0)	32261 (171,20)	32297 (32297,0)
32299 (173,15)	32303 (32303,0)	32309 (165,41)
32321 (162,59)	32323 (32323,0)	32327 (32327,0)
32341 (163,52)	32353 (32353,0)	32359 (161,74)
32363 (32363,0)	32369 (174,13)	32371 (161,75)
32377 (32377,0)	32381 (161,76)	32401 (161,80)
32411 (167,34)	32413 (32413,0)	32423 (32423,0)
32429 (175,11)	32441 (173,16)	32443 (32443,0)
32467 (32467,0)	32479 (176,9)	32491 (172,19)
32497 (32497,0)	32503 (32503,0)	32507 (32507,0)
32531 (164,49)	32533 (32533,0)	32537 (32537,0)
32561 (163,56)	32563 (32563,0)	32569 (179,3)
32573 (32573,0)	32579 (180,1)	32587 (32587,0)
32603 (32603,0)	32609 (162,67)	32611 (163,57)
32621 (169,29)	32633 (32633,0)	32647 (32647,0)
32653 (32653,0)	32687 (32687,0)	32693 (32693,0)
32707 (32707,0)	32713 (32713,0)	32717 (32717,0)
32719 (173,18)	32749 (163,60)	

Appendix C: A List of Primitive Roots

For each rational prime p, $p < 100$, the second column of the following table gives the nonassociated prime factor(s) of p, and the third column gives a primitive root of the nonassociated prime factor(s) of p.

p	prime factor(s) of p	primitive root of prime factor(s)
2	2	ω
3	3	ω
5	$2 + \omega$	2
7	7	$2 + \omega$
11	$3 + \omega, 3 + \bar{\omega}$	2
13	13	$3 + \omega$
17	17	$2 + \omega$
19	$4 + \omega, 4 + \bar{\omega}$	2
23	23	$2 + \omega$
29	$5 + \omega, 5 + \bar{\omega}$	2
31	$5 + 2\omega, 5 + 2\bar{\omega}$	3
37	37	$2 + \omega$
41	$6 + \omega, 6 + \bar{\omega}$	6
43	43	$2 + \omega$
47	47	$3 + \omega$
53	53	$2 + \omega$
59	$7 + 2\omega, 7 + 2\bar{\omega}$	2
61	$7 + 3\omega, 7 + 3\bar{\omega}$	2
67	67	2ω
71	$8 + \omega, 8 + \bar{\omega}$	7
73	73	$2 + \omega$
79	$8 + 3\omega, 8 + 3\bar{\omega}$	3
83	83	$2 + \omega$
89	$9 + \omega, 9 + \bar{\omega}$	3
97	97	$2 + \omega$

Bibliography

[1] G.D. Birkhoff and H.S.Vandiver, On the integral divisors of $a^n - b^n$, *Ann. of Math.* (2), 5, (1904), 173–180.

[2] Alfred Brosseau, *An Introduction to Fibonacci Discovery*, Fibonacci Association, Santa Clara, 1965.

[3] David M. Burton, *A First Course in Rings and Ideals*, Addison-Wesley, Reading, 1970.

[4] David M. Burton, *Elementary Number Theory*, Allyn and Bacon, Boston, 1976.

[5] R.D. Carmichael, On the numerical factors of the arithmetic forms $\alpha^n \pm \beta^n$, *Ann. of Math.* (2), 15, (1913/1914), 30–70.

[6] L.E. Dickson, *History of the Theory of Numbers* (3 vols.), Carnegie Institute of Washington, Washington, 1920 (reprint, Chelsea Pub. Co., New York, 1952).

[7] Harold M. Edwards, Fermat's last theorem, *Scientific American*, 239 (1978), October, 104–123.

[8] Harold M. Edwards, *Fermat's Last Theorem*, Springer-Verlag, New York, 1977.

[9] G.H. Hardy and E.M. Wright, *An Introduction to the Theory of Numbers*, 5th ed., Oxford University Press, London, 1979.

[10] I.N. Herstein, *Topics in Algebra*, 2nd ed., John Wiley and Sons, New York, 1975.

[11] J. Hunter, *Number Theory*, Oliver and Boyd, London, 1964.

[12] Dov Jarden, *Recurring Sequences*, 3rd ed., Riveon Lematematika, Jerusalem, 1973.

[13] Burton W. Jones, *The Arithmetic Theory of Quadratic Forms*, Mathematical Association of America, Washington, 1950.

[14] D.H. Lehmer, An extended theory of Lucas functions, *Ann. of Math.*, (2), 31, (1930), 419–448.

[15] Edouard Lucas, Théorie des fonctions numériques simplement périodiques, *Amer. J. Math.*, 1, (1878), 184–240.

[16] Ivan Niven and Herbert S. Zuckerman, *An Introduction to the Theory of Numbers*, 4th ed. John Wiley and Sons, New York, 1980.

[17] Harry Pollard and Harold G. Diamond, *The Theory of Algebraic Numbers*, 2nd ed., Mathematical Association of America, Washington, 1975.

[18] Leigh Wilber Reid, *The Elements of the Theory of Algebraic Numbers*, MacMillan, New York, 1910.

[19] Harold M. Stark, *An Introduction to Number Theory*, Markham Publishing Company, Chicago, 1970.

[20] Bryant Tuckerman, The 24th Mersenne prime, *Proc. Nat. Acad. Sci. U.S.A.*, 68, (1971), 2319–2320.

[21] N.N. Vorobyov, The Fibonacci numbers, in *Topics in Mathematics*, D.C. Heath and Company, Boston, 1963.

[22] A.E. Western, On Lucas's and Pepin's tests for the primeness of Mersenne's numbers, *J. London Math. Soc.*, 7 (1932), 130–137.

Index